Decision Engineering

Series Editor

Dr Rajkumar Roy
Department of Enterprise Integration
School of Industrial and Manufacturing Science
Cranfield University
Cranfield
Bedford
MK43 0AL
UK

Other titles published in this series

Cost Engineering in Practice
John McIlwraith

IPA – Concepts and Applications in Engineering
Jerzy Pokojski

Strategic Decision Making
Navneet Bhushan and Kanwal Rai

Product Lifecycle Management
John Stark

From Product Description to Cost: A Practical Approach
Volume 1: The Parametric Approach
Pierre Foussier

From Product Description to Cost: A Practical Approach
Volume 2: Building a Specific Model
Pierre Foussier

Composite Systems Decisions
Mark Sh. Levin

Yotaro Hatamura (Ed.)   25010|

Decision-Making in Engineering Design

Theory and Practice

Translated from the Japanese by Kenji Iino

With 250 Figures

 Springer

Yotaro Hatamura
Hatamura Institute for the
Advancement of Technology
Tokyo Opera City Tower 52F
3-20-2 Nishishinjuku
Shinjuku-ku, Tokyo 163-1452
Japan

and

Professor, Kogakuin University
Professor Emeritus, The University of Tokyo

Translator
Kenji Iino
Sydrose LP
475 North 1st Street
San Jose, CA 95112
USA

Originally published in Japanese as Koushite Kimeta by The Nikkan Kogyo Shimbun, Ltd., Tokyo, Japan

British Library Cataloguing in Publication Data
Decision-Making in engineering design. - (Decision
    engineering)
    1. Engineering design - Decision making
    I. Hatamura, Yotaro
    658.4'03'002462
ISBN-13: 9781846280009
ISBN-10: 1846280001

Library of Congress Control Number: 2005933049

Decision Engineering Series ISSN 1619-5736
ISBN-10: 1-84628-000-1   e-ISBN 1-84628-261-6   Printed on acid-free paper
ISBN-13: 978-1-84628-000-9

Printed in Germany

9 8 7 6 5 4 3 2 1

Springer Science+Business Media
springeronline.com

Preface

This book is a sequel to *The Practice of Machine Design*, and *The Practice of Machine Design, Book 3 – Learning from Failure*. It deals with what happens inside the human mind during such activities as design and production, and how we reach decisions.

Unlike other regular machine design textbooks or handbooks that describe how to accomplish good designs, the present volume explains what the designer thinks when making design decisions. A design starts with a vague concept and gradually takes shapes as it proceeds, and during this process the mind extracts elements and makes selections and decisions, the results expressed in sketches, drawings, or sentences. This book aims at exposing the reader to the processes of element extraction, selection, and decision-making through real-life examples.

Such a book has never been published before. An explicit description of the processes of making decisions, on the contrary, has been greatly needed by designers, and the managers of design groups have been much aware of such a lack. The non-existence of this type of book in the past is due to the following three reasons: the benefit of describing the mind process of design was never made clear, the method of such clarification was unknown, and no one ever invested the vast energy for producing such a manifestation.

Under these circumstances, we the members of the "Practice of Machine Design Research Group" boldly tackled the problem of expressing the decision processes in design and have documented our findings in this book.

This book first exposes what a "decision" is, what effect transferring its processes has, and how to realize such transfer. It then discloses some actual examples of decision-making processes in the design of manufacturing and production activities, and lastly explains the importance of grasping the process of the mind's decision-making by showing the reader an outstanding production system made possible through the disclosure of these processes.

People use their minds in performing activities. We make use of our brains especially when we take action or make plans or designs. The outcome of our minds is those actions, plans, or designs. The resulting plans and designs take physical forms, but what we thought about, what we wondered, the alternatives we tried, and what constraints we recognized while we were shaping our final results -

in other words, the scenarios - are usually not written up anywhere. Therefore, those who only get to see and read the results, *i.e.*, the final text and drawings, do not understand why the designer reached the conclusions, and they never learn the real design. To understand the real process is not just to map the results to one's mind, but to recognize that the entire scenario matches a template one already has in one's head. Everyone has templates in their mind in the form of scenarios, so what we want to learn also has to match such forms. Once we acknowledged these observations, we came to realize that expressing our brain activities when we design or handle manufacturing is most important in the real understanding of technology and related activities.

The authors believe that design is a typical field that involves creativity. The designer thus wants the necessary information, *i.e.*, what caused failures, how things were decided, and what plans were made. The designer wants to read his predecessor's account of "How I failed", "How I decided", and "How I planned". It was over 10 years ago that we made our plans to publish these books one by one.

The authors published the book in Japanese *The Practice of Machine Design, Book 3 – Learning from Failure* about 6 years before this book appeared there. The book was motivated by college students, who looked bored when they were listening to lectures about how to make things work, but displayed much enthusiasm when I started lecturing about my own failures. This made us realize that teaching the processes of failure has advantage in training the students about the real design. The book was accepted well beyond our expectation, and many people read it. We the authors, however, had our goal set on describing the mind processes of making decisions in words and illustrations. Unless this is done, we cannot understand the real problem.

After publishing *The Practice of Machine Design, Book 3 – Learning from Failure*, we immediately began work on this book. Just as with the former volume, 30 to 40 of us gathered, and each of us picked a topic we had made serious decisions on (the topic was a free choice) and started writing down the mind processes of those decisions. When we had our draft, about half of us had picked topics in design, but the other half had chosen to write about happenings in life when they were faced with hard decisions to make, *e.g.*, choosing a job, changing a job, divorce, and so on. The latter topics were not directly related to design. At that time, the authors were divided on whether to discard the second half and make the book deal solely with the topic of design, or to work on the entire world that surround engineers and discuss decisions we had taken that related to our lives. This debate cost us a delay for the fourth book[1] in "The Practice of Machine Design" series, in which we had been publishing a volume every Olympics year (every 4 years). Time passed without the authors reaching a consensus and we started to feel we might not be able to publish the fourth book. That would mean we could not publish the important book that would talk about the big theme of what we think when we make decisions! In 2001, Hatamura, who heads "The Practice of Machine Design Research Group," took the lead and restarted the

---

[1] This book is the fourth book in the original series "The Practice of Machine Design" in Japanese. The English series combined the first and second Japanese titles into a single book.

project of publishing this book with the policy that each author could pick any topic about any decision made in the past.

This book illustrates, with text and sketches, what people think and what decisions we make as we think. Anyone who reads and sees these texts and illustrations will learn about the mind activities of others, especially the detail of the "decision process." The reader will then probably receive some stimulation, realize the steps missing in their own decision-making, and apply their findings in making their decisions more fruitful, profound, and accurate. The reader will then acquire the skills of precisely transferring their own thoughts to the outside world. The world has so far not seen a book of this kind that directly expresses what goes on in our minds, and as this book is published, its contents are made to run on computers and become widely available in one form or another, a new field that makes "active use of the human mind" is opening in front of us.

The second half of the twentieth century was a time for us to develop our society by means of finding good methods established elsewhere, following them precisely, and producing mass quantities of the products quickly and inexpensively. The twenty-first century, in contrast, will probably further enrich our society by making direct use of the mental activities inside the human brain to generate "products," "systems," "methods," "sources of entertainment," and "roots that move our hearts."

Expressing the decision process of the mind takes text and graphic representation. The human mind makes various twists and turns which are hard to describe with a predetermined format or diagram. This book took up the challenge of presenting the mind activities using the following types of diagrams that we came up with.

- "Plane of thoughts diagram" that allows the writer to record whatever concept that arises in the mind without any organized format. Once these concepts are put on paper, the writer can usually see the relations among them, and generate the "Concept relation diagram" that shows the relations, sequences, and positions of these concepts.
- "Expansion of thoughts diagram" that expresses the sequence of high-level design from function to mechanism and to structure.
- "Spiral of thoughts diagram" that starts from vague concepts that capture the entire project, then gradually advances to details in a swirl.
- "Progress time history" that shows the time history of the process of events and related constraints.
- "Element relation spore diagram" (or "Mandala[2] diagram") that shows the group of structural elements by placing the high-level concept in the center surrounded by lower-level concepts and their relations to the central concept with lines.

The authors, through their work, developed these diagrams and we claim no best one among them. The writer will use whichever one, or number of them, is best suited for the situation at the time. It is impossible to describe the human mind process using just one type of diagram at all times. The human mind in fact, with

---

[2] Mandala: A Buddhist figure that symbolizes the awakening of the Buddha. It shows many small buddha's surrounding the Buddha.

the time axis, expands beyond three dimensions into the fourth dimension. Trying to express such a phenomenon forces us to use the diagrams listed above.

All articles about real-life decisions we made bear titles in active verb form using the first person. Through our third book, *The Practice of Machine Design, Book 3 – Learning from Failure*, we learned that noun phrase titles tend to a dull sound, whereas, verb phrases produce a sense of action and make searching through them much easier.

As regards structure, the book starts with four chapters that describe, what making decisions and transferring them mean, what decisions in design are made, how they are made, and what we need to know when making real products. The book then introduces decisions made by engineers in a variety of circumstances including production. In the last chapter, the book discusses the importance of disclosing the decision processes of the mind by providing an example that reviewed those processes to form connections with the production system, giving high efficiency and high precision.

Chapter 1 lays the foundation by describing what a decision is, the path to reaching one, how to express alternatives, and so on. The second chapter shows the need and actual methods for recording and transferring information about decision-making. It takes an example of "Reconstructed a hiking boot sole that came off during mountain climbing" to review the details of the decision-making process of the mind.

Chapter 3, "Decisions in Design," explains the meaning of decisions made during design, how they are made, and how to record the contents of a design. The chapter further lists the reasoning in decision-making and how to build an effective design support system. Chapter 4 contains a number of actual examples of the design process. It first shows design projects for university students: a hydraulic cylinder, a force sensor, and a positioning table. It then moves on to the design for building real-life projects with examples of a grinding tool that produces flat wafers, a micro-manufacturing system for mechanical processes under an electron beam microscope, and a controller for an automatic turbine blade grinding tool. The chapter also shows an example of building computer software through developing "Creative Design Engine" for aiding the designer in the conceptual stage of design.

Chapter 5, "Decisions for Production," explains some real decisions about technology development and engineering operations. For example, the description about technology development extracts the activities within the human mind, and categorizes them into groups of thinking, development, planning, improvement and research for disclosing the decision process. Chapter 6 highlights decisions about the relation between individuals and organizations. The chapter is partly about occupation, *e.g.*, changing jobs, starting new businesses, choosing majors in school, and relocation, and also talks about operation decisions like management, investment, human resources, and administration.

Chapter 7 discusses how to make use of the mind activity. It shows the construction of a new production system by first observing the decision processes during design and system architecture planning, which we used to bundle as part of the "human" area where we did not go into further detail. The chapter explains how the revolutionary metal mold processing system was created by INCS. It is a system with high efficiency, fast overall process time, and great accuracy. The chapter concludes that such disclosures of the decision processes will form the most important bases for the industry in the future.

Readers of this book, we assumed, are those who are starting on the design of real machines, having read *The Practice of Machine Design*, and *The Practice of Machine Design, Book 3 – Learning from Failure*. We, however, believe the broad topics we cover in this book will make it a good source of information for those in system design, marketing, and operations. In fact, we envision that the following types of people will read this book:

- First-time machine designers. Also, those who are creating large project plans for the first time, and novices who are not satisfied until they know reality.
- University (including college and graduate school) students who are learning design for the first time.
- University researchers planning projects that include designing and building experimental equipment for the first time.
- Engineers in design, research, or development groups who design or build systems and want to review their overall approach to projects.
- Those who are in a position to give instructions to students or group members, and have a strong desire to provide real-life knowledge and actual examples.
- Corporate managers who build product or manufacturing strategies and are about to restructure the overall organizational operation.
- Those interested in the mind processes of how people think and decide.
- Those who are at a loss, without confidence in their own ways of thinking or deciding.

For reading this book, we offer the following guidelines:

- Those interested in the decision process itself should read Chapters 1 and 2.
- Those who want to know about the processes of making design decisions and gain active use of them should read Chapters 3 and 7.
- Those interested in actual examples of machines and system design should proceed to Chapters 4 and 5.
- Those who want to learn about various decisions in real life should read Chapters 5 and 6.

Reading through the entire book will, of course, give the reader the whole picture of the decision process for design and manufacture, but selecting only the chapters to meet a particular interest will also provide the required information.

The discussion so far in this preface has dealt with our intention in editing this book. Hatamura, who heads our research group, wrote Chapters 1, 2 and 3. A few members of "The Practice of Machine Design Research Group" split the work of writing Chapter 4, "Examples of Design Decisions." The contents of Chapters 5 and 6 were contributed by the actual experiences of the group members. Here, we took the liberty of clouding some descriptions when accurate sentences might have caused trouble. We have not specified where we have used such techniques which were necessary to convey the real decision processes to the reader.

The following 51 people are the writers of this book. They are mostly graduates of our lab in the Mechanical Engineering Department of the University of Tokyo. All these graduates began their careers with the special design training our lab offered, which lasted over 30 years, and many of them entered manufacturing companies after they graduated. In a broad sense, they are all involved in design.

Yotaro Hatamura	Koji Ohgaki
Risuke Kubochi	Kozo Ono
Koji Yazawa	Koji Chikaishi
Norio Tezuka	Hideaki Yoshimatsu
Masaaki Iwasaki	Hideo Hara
Masumi Sekita	Masato Ihara
Shunsuke Kusama	Masao Fuchigami
Shinji Fujiwara	Kozo Sakai
Akio Inako	Kazutaka Hattori
Yasuo Shinohara	Takeshi Yoneyama
Toyotsugu Itoh	Eiji Mizutani
Hideo Hayashi	Tadayuki Hanamoto
Masayuki Nakao	Yoshio Yamamoto
Ryuji Takada	Kenji Iino
Tetsuhide Senta	Masataka Inagi
Shigeki Matsuoka	Tetsuya Hamaguchi
Kiyoshi Matsumoto	Kazuhiko Umezawa
Takao Odajima	Mitsuaki Adachi
Hirokuni Hachiuma	Katsunori Ichiki
Hiroyuki Sanguu	Kazuhisa Tanimoto
Shuji Tanaka	Kensuke Tsuchiya
Nobuaki Murakami	Masato Ohtsubo
Sota Fujioka	Tadashi Tsuji
Kei Nakata	

The following writers include a professor of our sister-lab, and wives of some of our members.

Takaaki Nagao	Masuko Tezuka
Junko Sekita	Tazuko Ihara

As we have said, this book addresses how our minds behave and how we reach decisions during one of the most wonderful and creative human activities, namely design. We have centered our discussions on systems and the design of machines. Our description of decision processes, not available anywhere before, will provide means for new manufacturing systems and production activities in the future. All of us will be more than happy if the book is found useful, as we have intended, in the practice of machine design and in enabling the reader to produce better designs.

Tokyo, Japan   Yotaro Hatamura
April 2005

Contents

1 What are Decisions? 1
    1.1   Understanding the Decision Process...1
    1.2   What It Is to "Make Decisions" 1
    1.3   Route to Decisions...5
    1.4   Decisions and Doubts 7

2 Describing and Transferring the Decision Process 13
    2.1   The Need to Know the Decision Process..................................13
    2.2   Method of Recording and Transferring the Decision Process 16
    2.3   Actual Methods for Transferring Decision Processes...........................20
    2.4   Diagrams for Expressing the Decision Process 22
    2.5   Example of Recording the Decision Process 27
            2.5.1 Describing What Happened 27
            2.5.2 Processes and Contents from This Event............................. 33

3 Decisions in Design 39
    3.1   Questions about Design 39
    3.2   The Mind's Decision Process During Design...........................40
    3.3   Constraints and Evaluation of Design 43
    3.4   Describing the Design Content...46
    3.5   Designers Want to Know the Reason for the Designs 47
    3.6   Design Support Systems the Designer Wants 49

4 Sample Decisions in Design..55
    4.1   Hydraulic Cylinder – the Mind Process of Designing One 56
    4.2   Designed a Torque Sensor 65
    4.3   Designed a Positioning Table 77
            4.3.1   Design Specification of the Positioning Table.......................... 77
            4.3.2   Analyzing the Functional Requirements.................................. 78
            4.3.3   Discussing and Determining the Basic Mechanism................. 78
            4.3.4   Producing the Structure 80
    4.4   Built an Intelligent Grinding Tool for Producing Flat Wafer Surfaces ..82

4.5   Planned and Proposed the Nanomanufacturing World Project 86
4.6   Developed the Control System for Automatic Grinding of Turbine
        Blades 90
4.7   Built "Creative Design Engine" for Assisting Idea Generation 98
4.8   Guided Students' Free Imagination in Building Stirling Engines 102

5 **Real Decisions in Manufacturing** **113**
        5.1   Decision-Making in Technology Development 114
                5.1.1 Finding Out the Real Safety System at Mt Usu –
                        Thinking 1 114
                5.1.2 Designed a Pressure Sensor Exposed to Severe
                        Conditions – Thinking 2 118
                5.1.3 Reduced the Weight of an Automobile Compressor –
                        Thinking 3 122
                5.1.4 Automated the Narita Express Car Junction Hood –
                        Thinking 4 127
                5.1.5 Developed a Telescopic Arm Clamshell Digger –
                        Development 1 131
                5.1.6 Developed an Automatic Segment Construction Robot for
                        Sealed Tunnels – Development 2 137
                5.1.7 Developed a System for Preventing Mobile Crane
                        Overturn – Development 3 142
                5.1.8 Modified the Lighting Unit of a Wafer Character
                        Recognition Machine   Practice 1 148
                5.1.9 Succeeded in Laser Welding by Controlling Its Tip
                        Distance – Practice 2 154
                5.1.10 Applied IC Tagging for Managing Metal Mold Parts –
                        Practice 3 158
                5.1.11 Modified a Sandblast Machine for PDP Class Substrate
                        Machining – Practice 4 161
                5.1.12 Installed a New Cable on a Multi-Joint Robot –
                        Modification 1 164
                5.1.13 Modified a Material Cooling System but Could Not Cut
                        the Cost – Modification 2 170
                5.1.14 Successfully Anchored a Floating DNA Fiber in Liquid
                        – Research 1 171
                5.1.15 Made DNA with Fluorescent Molecule Visible in Liquid
                        – Research 2 176
                5.1.16 Built a Microscopic Assembly Tool – Research 3 180
                5.1.17 Accomplished Wide-Range High-Precision Positioning
                        – Research 4 185
        5.2   Decision-Making in Technology Management 189
                5.2.1 Selected a 3D CAD System – System Introduction 1 189
                5.2.2 Arbitrarily Selected a CAM System – System
                        Introduction 2 192
                5.2.3 Evaluated a Technology for Breaking Rocks with
                        Electromagnetic Force   – Technology Introduction 1 195

5.2.4 To Introduce a New Forming Method to Our Main
Factory or Not – Practical Solution 1 198

6   Decisions about Individuals and Organizations............................203
6.1   Decisions about Occupations...204
6.1.1 Jumped to a Small Company for a New World – Job
Change 1.. 204
6.1.2 Moved from Manufacturing to Consulting – Job Change
2... 207
6.1.3 Moved from a Company to School for a Diploma – Job
Change 3.. 210
6.1.4 Resigned from a Trading Company and Became
Independent – Entrepreneurship 1 212
6.1.5 Started a New Business in the US for Building a Bridge
over the Pacific – Entrepreneurship 2.................................... 218
6.1.6   I Became Disabled and Selected the Course of My Life
– Turning 1 221
6.1.7   Jumped into an Unexplored Research Area – Turning 2......... 225
6.1.8   Built a New Lab and Research Group in a Conventional
Field – Turning 3 227
6.2   Decisions about Corporate Management 231
6.2.1 My Company will Disappear in One and a Half Years –
Operation 1 231
6.2.2 Forced an Organizational Change in a Traditional
University   – Operation 1 233
6.2.3 Located a New Factory – Investment 236
6.2.4 Selected My Successor and Recommended Him –
Resources 1... 239
6.2.5 Restructured a Department with My Own Thoughts and
Failed – Resources 2.. 241
6.2.6 Started a Young Engineers Training Program –
Management 1 243
6.2.7 Applied to a MITI Project but I Was Declined –
Management 2 246

Applying the Mind Activity for Manufacturing............................249
7.1 Where in Manufacturing to Apply the Mind Activity 249
7.2 Decisions by INCS..250
7.3 Implementation by INCS 252
7.4 Outcome of the INCS Development 256
7.5 Where Does This Lead Us to in the Future? 257

Postscript...261

Index 263

1

What are Decisions?

1.1 Understanding the Decision Process

When we engage in the actual work of design, we often doubt if the way we make our decisions is acceptable or if there are any better solutions. Examining past designs by others also leads us to questions like why this object is built that way or how the designer worked out the structure. When you are a novice designer, or even if you are an experienced one boldly entering a design field new to you, you have no idea where to start, what parameters are fixed, which are free for you to choose, and in what sequence you can determine them.

All these questions relate to the mental process during design, and especially to the process of making decisions. The designer has to do the following for himself:

- Understand how the mind makes judgments, where such processes are positioned, and how to proceed from where he is.
- By knowing how others made decisions, understand the overall picture of applicable technology and clarify its relation with the designer's own judgments (Chapter 3 goes into the details of this process, which one can only carry out through the transfer of technology).
- Grasp the overall picture of the design process, that is the sequence of determining matters including what the goal (functional requirement) and conditions (constraints) are, so that he can apply them to his own activities (design).

1.2 What It Is to "Make Decisions"

What, then, does it mean to "make a decision"? Instead of a complex, meaningless definition, I would state it as follows:

When there are multiple options that could happen or one could make happen, think about the probability for each option and set one of the probabilities to 1 (100%) and the rest to 0 (Figure 1.1(a)).

In other words, "making a decision" means to select one from multiple choices.

If we look at "decision-making" from another angle, of change of state, "to decide" is the transition of the mind potential (I am not sure if such a phrase exists) from one state to another through an active move by the person making the decision (Figure 1.1(b)).

The reason for such roundabout expressions is for the reader to better understand the creative mind decision processes like "selection", "hesitation", "inertia of the mind", and "obstacle of the mind" that they will be later exposed to.

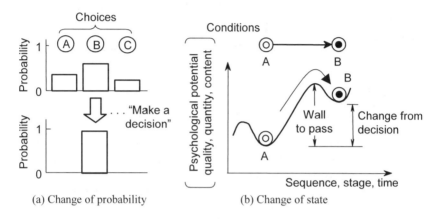

Figure 1.1   Making a decision

For now, let's say to "decide" is to "select" and move on to the question of what types of decisions there are. Decisions, most likely, can be grouped into three types; (a) go or no go, (b) single selection, and (c) structured decision. The (a) go or no go type means there was only one choice to begin with and the decision is whether or not to do it. The second type, (b) single selection, indicates the situation at a single node with multiple choices, from which one is selected, and (c) is the state with multiple nodes, each with multiple choices, and selecting one option at each node leads to a structured route. The designs this book introduces are rarely of the second type and mostly form the later, structural selections. Through this book, we want to disclose the selection structure to the reader so that they understand that the contents of design mean nothing other than to learn the selection structure.

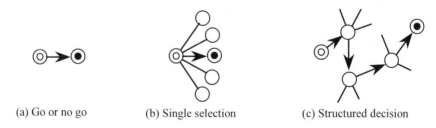

Figure 1.2   Types of decisions

So how are decisions made? Figure 1.3 shows the process. We do not freely make decisions, but are always under some constraints (Figure 1.3(a)). In this situation, making a selection is quite complex. When we have some method of evaluation, the process is easy to follow even for an outsider, but it is an act by a live human, who is often affected by their own experience, habit, preference, and mood at the time. We, nonetheless, do not rely entirely on chance every time we make a selection. The actual decision process follows Figure 1.3(b). The figure shows our mind process of constructing a virtual structure for even the simplest selection. The process categorizes the choices, reconstructs them, and evaluates each option by some method. If the terminal node does not determine our choice, our mind goes back up the structure to reevaluate other options, and we go back even further if we are still undetermined.

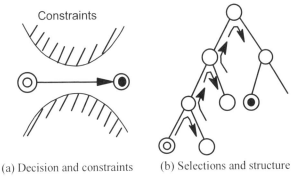

(a) Decision and constraints   (b) Selections and structure

Figure 1.3   Decisions constraints and selections

The situation is even more complex in the case of structural selection (Figure 1.2(c)). In this case, we set the goal as Figure 1.4(a) shows and come up with a route from the start point (the first point) to the end point (goal). For example, the start point is a function and the end point a mechanism, or the start point is the shape and the end point is how to manufacture it. The complete route we construct, as Figure 1.4(b) shows, is a passage through structural elements. The actual mind activities that take place within the human brain, however, go through many searches and possible selections, as Figure 1.5(a) shows. Figure 1.5(b) gives a simple idea of these searches; however, the reality is more complex, irrational, and changes dynamically with time.

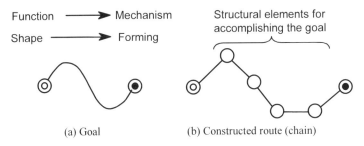

(a) Goal   (b) Constructed route (chain)

Figure 1.4   Route to decisions

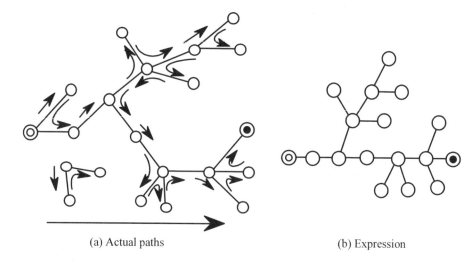

(a) Actual paths   (b) Expression

Figure 1.5   Expressions of routes to decisions

Many believe that we make selections based on certain evaluation criteria like experience, preference, or the text of manuals; however, as the science of the brain develops and uncovers the actual mechanism of making selections, it may show that the mechanism is totally irrational from our own standpoint. That is, it may not take the form of an input from the left entering the operator element called "evaluation" which then spits out the resulting selection as output from the right; instead, the decision-maker may just make selections based on whether the choice matches past experience of success or failure. Such experiences are totally individual. Or the decision-maker may construct a successful route within the mind first and then search the field until a route, starting from the right, is discovered that links to the input at the left end. This is a reverse operation, in other words, "selection through hypothesis and proof." Sometimes the decision-maker is not aware of using this technique; and seeing such quick decisions made, we often wonder about an invisible mysterious circuit and call it a hunch. In any case, studies on the science of the brain are advancing at an enormous speed and we may soon have to rewrite this entire section about making selections!

1.3 Route to Decisions

We make decisions along the routes of thoughts. How are these routes determined? As I explained in the previous section, it seems that they are not successively set from left to right, as for example Figure 1.5(a) and (b) shows. So, how are they determined? Figure 1.6 gives the answer. When our mind is given the starting point (proposition or problem) and the end point (goal or solution), it then, all at once, brings up anything that may relate to them, and without any particular relations between the concepts. We then lay out these related concepts on a plane to expand the thoughts (let us call it the "plane of thoughts") in a random manner, a process we call "isolated scattered placement."

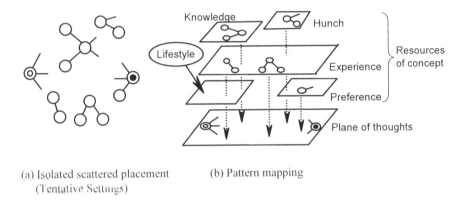

(a) Isolated scattered placement   (b) Pattern mapping
(Tentative Settings)

Figure 1.6   Basic process of decision-making in the human mind

These fragments of thoughts that arise from our personal resources like experience, hunch, preference, or knowledge are mapped to the plane of thoughts all at once. Among these pieces, those that contribute the most to our process are those based on individual experiences. The role of knowledge is relatively small compared with that of experience, which is why people with ample knowledge but hardly any experience cannot make decisions when they are really needed. The mind then partially connects these fragments into lumps of thoughts with logical reasoning (call them clusters), then connects these lumps together to construct a network of thoughts. It then selects some routes from the network and based on its own schemes of evaluation (matching with experience, preference, knowledge, or benefit) sets the final route.

This decision process is different for each individual. Therefore, just hearing the details of such processes made by others leaves isolated points that make the understanding of the whole picture difficult. The diagram in Figure 1.5(b) is an expression that makes the understanding and transfer to others easy with its description flowing from left to right.

Let us now look at how the routes of thinking grow and decay. Frequent stimulation to the neurons and synapses in the human brain makes them grow larger, letting signals that pass through them travel faster. This is what "learning" and "practice" are. The routes of thinking undergo these transitions, as Figure 1.7 shows. Learning makes particular routes of thinking larger and generates more branches of choices. Repeated thinking about the subject and having a high level of curiosity are important for keeping the possible routes in good shapes - a state that allows proper handling of unexpected situations and adequate thinking with plenty of margin. If the route is always the same without any new stimulation, the number of options decreases and the speed of thinking also declines. When an unexpected obstruction interferes with the route in such a state, the network cannot handle the disturbance and disintegrates. This diagram teaches us the importance of study and practice as well as the role of curiosity, what the aging of humans and of organizations means, and so on.

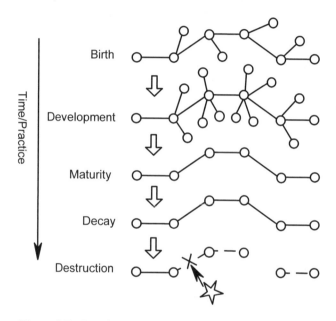

Figure 1.7   Development and decay of the route of thinking

Next, we may further study the need for experience and knowledge. Experience and knowledge acquired through solid recognition allow complex formation of one's network of thoughts. A route that only has poor choices in Figure 1.8 is thin, weak, and disintegrates when just a single disturbance breaks a single spot. The small break then propagates through the entire passage from the start to the end. On the other hand, a network of rich choices has many branches that spread the space, in a profound and ample manner, and is solid. The start and end points are connected by the selected route, as well as other routes, giving the network the strength to reconfigure itself in case of unexpected external disturbance that cuts a path in the solution. The route of decision in such a network is "rich," "modest," in the sense that it does not excessively exaggerate itself to the outside, "flexible," in that it easily reconfigures itself against changes in the outside world, and is eventually marked as the "proper" route.

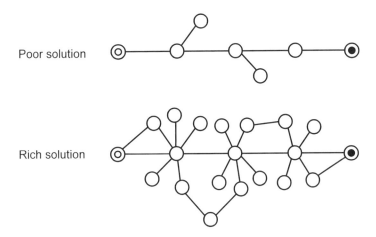

Poor solution

Rich solution

Figure 1.8   Poor and rich solutions

For immediate reconfiguration to take place, just having a large selection of routes is insufficient, and the paths of the network need to always be alive (a state we can call one of "active paths"). Keeping the paths active requires actually using them occasionally as well as practicing "virtual exercised" in which we think of how to apply the routes hard and repeatedly.

Here is the relation between "creation" and our networks. I earlier explained that making a decision is marking the candidate routes and selecting the final ones. Note that the passage is not constructed through logical thinking or deduction, as is generally believed. Fitting the candidate elements into a structured logic for evaluation may be a form of hypothesis and proof. Logic, however, cannot help finding which element to fit to which part of the structure and how to evaluate them. Logic and deduction can only help in evaluating an already constructed passage and not in "how" to construct the passage itself. What provides help in constructing the passage are personal experience and feelings accumulated through real experience and knowledge.

1.4 Decisions and Doubts

When we make decisions and even after we have made them, we always doubt if we will make or have made the right choice: "Which one should I pick?", "Is this the right one?", "Was it right?" Such doubts always accompany decisions. It always happens and the two cannot be separated. What makes us doubt our decisions? Let's go back to Figure 1.2 and discuss the matter.

First, for go or no go decisions: There is only one choice in this case and the selection is whether to go or not to go. The doubt here is not about our selection, but it is whether we can get over the hill of psychological potential in Figure 1.1(b) or not. The state in which we have not climbed the hill is called "hesitation," and occurs especially when we hesitate because we are not well prepared for the consequences to follow we call the state "indecisive." Indecisiveness is the worst state in making a decision. The loss of time and opportunity can be fatal. It is a state that we have to avoid both in our individual lives and when we are running organizations. We have always to keep striving to acquire experience and knowledge, and carrying out virtual exercises.

Next, for single selections: what gives doubt here is the fear that we may have not listed all the possible choices, and, not having assigned the probability for each choice, we cannot reach a final verdict on which option to back 100%. We will always have doubts when we come to these situations, and they are not necessarily something to avoid. To make our selection quickly and correctly, however, we have to keep up the experience, knowledge, and virtual exercises in this case as well.

There are situations in which the project leader continues to have a "slight sense of anxiety" even when he has already set the overall plans and the project has already started. Such anxiety is caused by the persistence of the "doubts" the leader had while having to assign a 100% probability to one choice, even when there was uncertainty. Leaders are always required to bear with this burden.

In 2002, Super-Kamiokande (an observatory of neutrinos and other cosmic rays built where the Gifu-prefecture Kamioka mining facilities used to be) experienced a chain explosion of photomultiplier tubes, which sense cosmic rays, and the project leader had some anxiety about an accident from unknown causes at the time (Lecture by Professor Yoji Totsuka of the University of Tokyo, March 8, 2003).

Lastly, about structural selections: this type of doubt is natural in making decisions during the construction of a structured solution. Figure 1.9 shows the selection and doubts in such decision-makings. Figure 1.9(a) shows the selection of routes for a decision. We gather cases of isolated scattered positions shown as in Figure 1.6 and, for the passage we constructed, wonder if the sequences of routes were adequate or if we should have taken different routes. This indecisiveness originates from our selections at each branch. The source of our doubts in this case is our fear lest the final passage from the start to the end is a sequence of local optimums and lest the solution may not be the overall optimum.

Figure 1.9(b) shows an analogy with mountain heights. When we want to climb the highest mountain, we cannot be certain which strategy is the best: (A) to follow the locally highest points towards the top, (B) to temporarily descend to a low point but then to come up to a higher point than by following just the peaks, or (C) if there is a single highest peak irrelevant to all the lower peaks altogether. Such doubts are natural and to truly reach highest peak we need a bird's-eye view of the region around the mountain. This means to have all the experience and knowledge about the main elements that relate to the mountain. In addition, carrying out virtual exercises and research on a daily basis can ease the anxiety associated with doubts of this type.

One thing about "doubts" we need to be aware of is the relation between "route of decision" and "psychological obstacles." We often face the situation where we could only reach the solution next to the best due to "psychological obstacles" caused by too much experience, insufficient knowledge, the air at the time of making the decision and other factors. This always happens with human activities and decision-making, and it is the source of our doubts when we select structures. To overcome these obstacles and reach the optimum solution, we need to be aware of the persisting presence of these obstacles and carry out plenty of virtual exercises based on experience and knowledge.

One technique for overcoming these "psychological obstacles" is seeking advice from others at the same level of information and mind potential, or brainstorming. TRIZ is one of the scientific methods invented in Russia 50 years ago by analyzing how the human mind overcomes these psychological obstacles. It is highly effective for creating route for thoughts.

Conveying the decision-making process with sketches of mountain climbing worked very well. Figure 1.10 compares the case of continuous endeavor and relying on mediocre judgments with corporate decisions and executions. When we need to decide whether to take the right or left route, those who can foresee continuous hardships on the proper route often end up selecting the other one. Once we select such an easy route, we have to keep walking along it until we are faced with destruction. Though we may realize the mistake midway, the ridges are too high for us to switch over to the proper route. The hiding of recall cases by Mitsubishi Motors in 2000 was a typical case of the wrong choice at such a juncture.

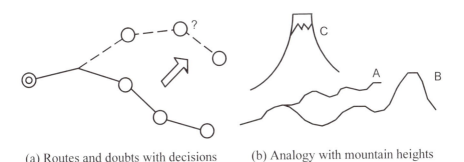

(a) Routes and doubts with decisions   (b) Analogy with mountain heights

Figure 1.9   Selections and doubts with decision-making

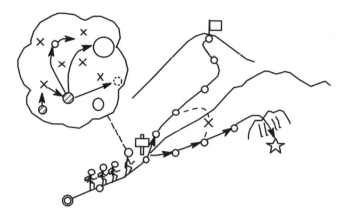

Figure 1.10   Virtual exercise determines everything – the highest qualification required of the leader

Everyone has his own psychological obstacles, and they do not always bring negative results. As long as we stay within the bounds of the obstacles, we can reach solutions close to the correct one "skillfully and swiftly." The real solution, however, is often on the other side of the wall. TRIZ is one of the methods for overcoming these obstacles. The way it works is equivalent to what we may call the "principle of creative design." The principle of creative design aims at applying knowledge and common sense from other fields of study to our own field so as to overcome psychological obstacles.

The failure of the H2 rocket launch in 2000 was due to gas leakage from the combustion chamber affecting the control system situated next to the leakage location. The chief designers of automobiles already have the common sense to avoid such an arrangement. It is a desirable practice for engineers to actively acquire the common sense of other fields.

Note the following about the relation between creation and doubt. Creation always proceeds through hypothesis and proof, therefore a creation is always accompanied by doubts. Doubts that are always present to the end include whether the assumptions were sufficient and adequate, if the proof was correct, or if the selection itself was correct. Time passes and the situation changes while we do not reach reasonable solutions and we have to make the next decision. We will understand that "creation" and "doubt" are two concepts that always go together and learn to deal with them appropriately.

Why Did Diamond Princess Burn?

In the evening of October 1, 2002, the soon-to-be world's largest luxury passenger ship *Diamond Princess*, under construction in Mitsubishi Heavy Industry Nagasaki Shipyard, had a fire which continued to burn for one day. The huge vessel would have weighed 113,000 tons, with a length of 290m, a width of 37.5m, and a height, from the keel, of 62m. The ship was carrying Japan's hope of a new direction for the future of the shipbuilding industry.

As I watched the fire throw up black smoke, many questions came to my mind. Why did the fire start? The builders must have been well prepared in fire prevention and fire fighting training; did these not work or function to plan? Had a malicious arsonist selected the weakest point for putting out the fire and at the worst time?

After I had pondered a while, mind settled on the following two thoughts: One is that all fire prevention and fire fighting systems must have been planned in the forward direction. Maybe there was no reverse thinking of what series of events (scenario) would cause the ship to burn down. In other words, there was no total system evaluation from the viewpoint of the "devil of failure with ill intention." Another thought was that maybe the person who first discovered the fire had never had the experience of seeing a real fire, despite of all the fire fighting training he must have had. And when he saw the real fire, his mind "blanked out" and his body just stood still even if he knew what he had to do.

This chain of thoughts is detailed in our previous textbook *The Practice of Machine Design, Book 3* and my book *Learning from Failure* published by SYDROSE LP. It must have been God's will that the fire started in the same facilities next to the turbine rotor destruction case in the former book. Curiously, this fire broke out 32 years after the turbine rotor accident. It just matches the "30 year cycle of accidents" theory in the study of failure. Maybe there is some invisible force present.

2

Describing and Transferring the Decision Process

Chapter 1 showed what a decision is. This chapter discusses the significance, for general engineering, of describing the decision process to others and what we need to do to successfully have others understand the decision process we have pursued.

2.1 The Need to Know the Decision Process

Why is it necessary to describe the decision process of inventing a technology when the technology already exists? Figure 2.1 shows the reason. Just looking at an established technology does not allow us to reach a real understanding of it, to use it and to develop it further. To enable such further steps, we need to see the process of giving birth to the technology, especially the process of making decisions. We need to know the mind process of inventing the technology (Figure 2.1(a)), and this means the inventor has to record the mind process in a form that can be communicated to other people. The description of such records has to exceed the threshold necessary for technology transfer (Figure 2.1(b)). Once the description surpasses the threshold, the technology can transfer over time, space, organization, culture, and technology field; however, if the record level remains below the threshold, the technology just disappears.

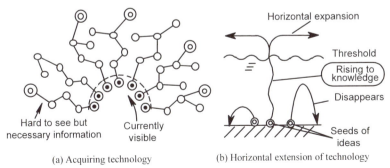

(a) Acquiring technology   (b) Horizontal extension of technology

Figure 2.1   Why we make records of decision processes

13

Figure 2.2 shows what happens when we use records of decision processes. These situations include the following:

- When the use is undetermined at the time of making the records but the individual or the organization requires them.
- Supporting the processes of thinking, making decisions, and performing virtual exercises by technology creators.
- Supporting mutual understanding among technology creators in the group.
- Transferring and understanding technology created at or in different times, spaces, organizations, or cultures.
- Education for people to create future technology.
- Constructing systems that support the thinking processes of technology creators.
- Grasping the overall picture of design processes. The picture includes the sequence of determining matters including what the goal (functional requirement) and conditions (constraints) are, so that the designer can apply them to his own activities (design).

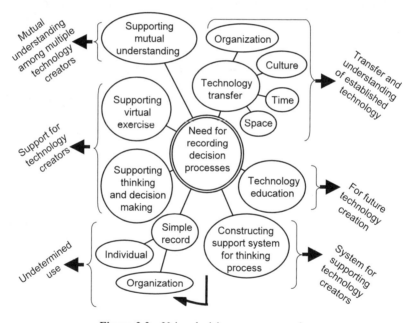

Figure 2.2   Using decision process records

Transferring the decision process is part of the wider task of "technology transfer." Figure 2.3 shows the stages that require technology transfer. They reduce to the following three situations:

(a) Across time, generations, and people:

- Different people separated by a time gap: This is what we call passing technology across generations. Many of old technologies are still needed today and unless this type of technology transfer is properly made, the whole society that relies on technology disintegrates.

- The same person at different times: We know the difficulty of remembering everything we have thought about in the past. We should occasionally review our own thoughts and actions from the past.

    Our own memos, sketches, or sentences recorded months or years ago often surprise us. I am often amazed by "That's what I was thinking!" or "What a surprise that I had accurately foreseen the situation now!" My record-keeping of actual decision processes gives me such experiences. We should always date and place titles on our records and construct our own "idea notes." I have had regrettable experiences from not marking these two even though I had the sketch and sentences. Remember to bear in mind the importance of date and title. We can also say that a memo without the date and title is the same as not having one.

- Across generations: People across different generations, even when they exist at the same time, have different experience and ways of thinking. Even though they are at the same place at the same time, they need to consciously transfer the technology to be successful.

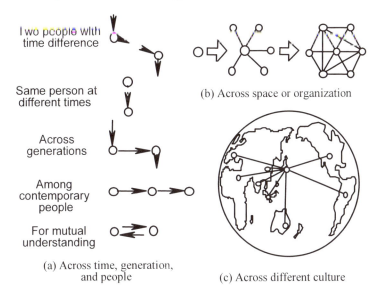

Two people with time difference

Same person at different times

Across generations

Among contemporary people

For mutual understanding

(a) Across time, generation, and people

(b) Across space or organization

(c) Across different culture

Figure 2.3 Where technology transfer is needed

- Among people at the same time: As information travels, the original contents decay over distance and they often get exaggerated and twisted. To avoid such mistakes, we need to give a solid structure to the contents.

- Mutual understanding: Concepts only transfer in a form acceptable by the receiver. We need to arrange the structure so that the receiver can share the contents.

(b) Across space or organization: Technology first starts spreading where it was born. Then its derivatives start to relate to each other in a network. When the organization that holds the technology spreads in space, the technology that originated there also spreads in a radial direction and then forms a network. The technology itself needs a solid structure as it is subject to transitions through related technology and organization.

(c) Across different cultures: Different cultures think and accept technology differently, and sometimes even evaluate it differently. We have to structure the technology from the beginning so that its records can survive such differences.

So far, we have looked at situations where we need technology transfer. For technology to transfer in the real sense, the mind process of the person who made the technology needs to transfer to the receiver. The receiver wants to understand the mind process of what, why, and how decisions were reached, and then what followed, what lessons were learned through the decision and how to make use of the decisions. The real technology transfer is made only when this information is successfully passed on to the receiver.

Many Japanese corporations are shifting their production facilities to China. Transferring production by just sending the machines to new locations does not work, and neither do the blueprints. What can make the transition easier? This book has the answer. It is to transfer the mind process of making decisions. The effectiveness and speed of absorption are high when the information originator provides what people are looking for. Japan is at a crossroads between following this book and transferring the mind processes to foreign countries or to hiding them away in black boxes. After all, the only way for Japan to survive is to place its own technology development in a black box and prevent it from leaking out to others.

We, however, shall bear in mind that the development of technology has to obey the legal system and artificially twisting the law only has temporary effects. Even if we hide the mind's process of decision-making, others will sooner or later develop their own. The advantage of hiding the processes will last only for a short time. See Korea's Pohang steel factory. They built it on their own while Japan hardly provided any help, and it is now the world's largest steel manufacturing facility.

2.2 Method of Recording and Transferring the Decision Process

How can we make event records that easily transfer the decision processes to others? "The Practice of Machine Design Research Group," which is the group that is authoring this book, have long been discussing this topic. One outcome of our discussion is the book *The Practice of Machine Design, book 3 – Learning from Failure*. In this book, we broke down a failure event description into 6 topics of event, process, cause, action, summary, and knowledge. This format allowed uniform handling of failure cases and we were then able to widely transfer the knowledge. We learned through this experience that for describing complex matters like how the human brain operates, we need a structure for organizing the information so that it flows in the natural way of the human mind process.

We then advanced our discussion into the topic of how to describe the decision process. It was then that we found that the topics in Figure 2.4 are needed for recording what, why, and how decisions were reached, what followed, what was learned through a decision and what to do to make use of it.

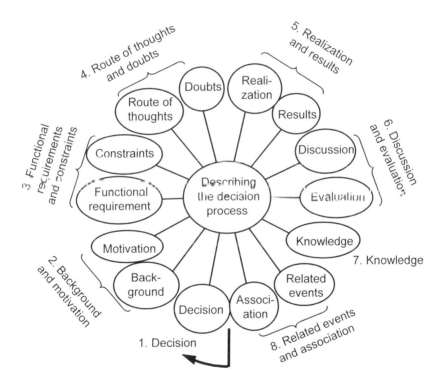

Figure 2.4 Topics necessary for transferring the decision process

The eight topics in the figure are:
(1) Decision: Record what decision was made.

(2) Background and motivation: Circumstances and supplemental information at the time of making the decision. Also record what made the decision necessary.

(3) Functional requirements and constraints: What functional requirements and specification the decision had. Which constraints were clear at the time of making the decision, and which were hidden then but were revealed later.

(4) Route of thoughts and doubts: Show the route the mind took from the problem statement (starting point) to the results (end point), what options were available for each selection, and what doubts entered the mind.

(5) Realization and results: Record the time history of events upon executing the decisions and what results followed.

(6) Discussion and evaluation: Looking back at the decision, its realization, and results, what should and should not have been done. How to make the decision useful and how some of the points are evaluated.

(7) Knowledge: What was learned as knowledge through making the decision and what followed.

(8) Related events and association: Events that resulted from the decision (direct results), and events that took place in relation to the decision (indirect results), and other thoughts that enter the mind in relation to the decision.

Among these topics, route of thoughts and doubts, discussion and evaluation, knowledge, and related events and association are usually not recorded. Technology, however, is only transferable in the real sense by recording these three topics and only then the "decision" by a single person turns into wealth shared by other people and this is what we call culture.

Once we have these descriptions, how do we use them? Figure 2.5 shows how decision process information transfers in terms of technology transfer.

- Record: Write a description and record it. At the same time separate the writing into the above topics of decision, background and motivation, functional requirement and constraints, route of thoughts and doubts, realization and results, discussion and evaluation, knowledge, related events and association, and so on.

- Store and archive: Whether we will use the information or not, we will keep a record in the form of text, figures, animation, and sound, and in addition we may want to keep the real object we are making a record of or others that caused it. Make sure to keep the records in a dynamic area that we visit often instead of storing them in a stationary manner. Unless we make these types of storage in 3D, mere documents and diagrams can never really transfer technology. Also, simple records without a scenario will never be used in later days. Just making records is the least work necessary; however, it is not sufficient. Whether it is in text or a photograph, the "scenario" we made at the time of record-making sets the value of the data.

- Search: Analyzing the contents and giving them a structure allows making use of the contents. Mere description and storage without consideration of the use are equivalent to having nothing.

- Distribute: After analyzing the contents, distribute them to proper areas of technology, product, and department so that the technology spreads into the network. This idea comes from Mr Fukuda of Mitsubishi Heavy Industries Nagasaki Shipbuilding plant in the turbine rotor destruction article in *The Practice of Machine Design, Book 3 – Learning from Failure*.

- Publicize: Keeping the description in a dead storage is pointless. To make it common knowledge for all, we must place it in the media like books, magazines, newspaper, television, and computer networks as well as distributing the information at conferences, on journals and via patent announcements.
- Educate: Turn the described contents into knowledge and transfer them through education with real experience or virtual ones to the people of the next generation.
- Make business: Think about turning the contents of the record into a business.
- Socialize: Try to have the contents socially accepted, obtain intellectual properties, and acquire legal responsibility like product liability.

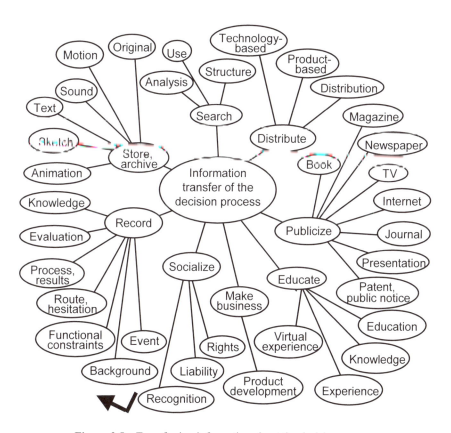

Figure 2.5 Transferring information about the decision process

2.3 Actual Methods for Transferring Decision Processes

If we really want to understand the contents of a decision that someone else has made, reading only about the final decision does not take us anywhere. Information from just the decision itself does not let people understand what facts formed the basis for the decision, what and how the decision-maker thought and what he did (Figure 2.6). That is because the decision process proceeds, as the figure shows, from the left to the right following the time line and those only at the right end cannot trace the time backwards. In other words, the readers at a later time cannot understand the actual essence of a decision made earlier due to the "irreversibility of time"; it is as if there is an invisible wall.

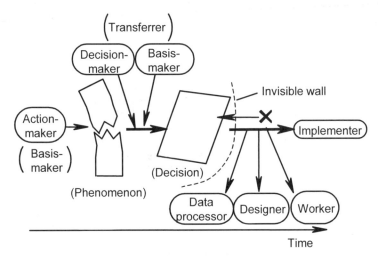

Figure 2.6. Those who implement a decision often do not understand the basis of the decision nor the decision-maker's mind

If someone was shown just the final decision and was told to make exactly what he sees, would he be able to do so? Figure 2.7 shows how a mountain hiking boot was repaired. This sketch alone cannot show much of the repair. Section 2.5 later explains this topic in detail.

Thinking that a blueprint alone is sufficient for building a part is the same as assuming that one can set the string tension by looking at just Figure 2.7.

So, is it enough for real understanding to disclose the cause that led to the decision together with the final results? Figure 2.8 shows the original problem with the mountain hiking boot. The figure reveals more information than the previous one; however, in terms of transferring information to the reader it fails.

It is for the same reason that people do not make good use of accident reports. Even if the writer adds a description of the cause to the results and actions, the reader cannot plan to carry extra strings to tie up the broken hiking boot.

Why are such descriptions insufficient in transferring information? It is because they do not contain all the mind processes of the person who had the experience;

only the objective information. They are also written from a third person's perspective. Chapter 1 explained these mind processes – the wonders, trials, background, motivation, and what came up in the mind of the person with the experience. Real information transfer requires all the mind processes of the person from the first person's perspective.

Figure 2.7   Only showing the decision transfers little information

(a) How the sole came off   (b) How the boot was eventually tied up

Figure 2.8   The cause and results alone do not transfer much information

There are many trouble and accident reports, but no matter how hard the editors try, people hardly read them because they do not lead to a real understanding of the events, and thus people cannot make use of the information. A trouble report that contains only the event, cause, and action fails to transfer the mind process of those involved in the event to the reader who is trying to relive the event. It is for the same reason that a blueprint alone cannot transfer technology.

For real transfer, we need to describe the mind process that takes place within our brains. Figure 2.9 shows an example. Once a problem is set, we think and execute much to solve the problem. Record what enters to the mind, how you wondered, what you tried, and how it all came out. Then describe what value judgment you made and how you decided about the solution. This series of writings records the decision process, but it is also important to add the results of a virtual exercise to include what came to mind in relation to the decision, what effects the decision may cause and what are the countermeasures.

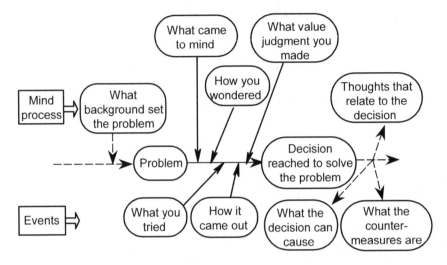

Figure 2.9   Describing the mind process from when the problem is stated

2.4 Diagrams for Expressing the Decision Process

There are two ways of expressing what goes on inside the human brain; one is by words and sentences, and the other is by diagrams. Humans tend to mix the two expressions, and deep within the mind, concepts, which are not in either form, move around. If we force ourselves to express them, we end up using words and diagrams.

People that are literature-oriented like to express thoughts with words, whereas those more science-oriented feel diagrams are more precise. There are a few that like to use equations and we may want to call them "math-oriented people".

Professor Ikujiro Nonaka claims it is important to express "thoughts by words, and words by shapes", but here we assume words and shapes coexist without either being superior to the other.

Now let's think about what type of diagrams we should use for expressing the thinking process. The author group concluded, after six years of discussion and trial and error, that the following diagrams are effective.

(a) Plane of thoughts diagram (Figure 2.10): Diagram that records words and pictures (we will collectively call them "concepts") in a random manner as they enter the mind. The words can be nouns or verbs, as well as numbers with units. We do not have to stick to words, and can draw pictures as well. Concepts in the mind are not always suited to expression by words, and pictures or combinations of the two types may well describe them. We can freely use these forms.

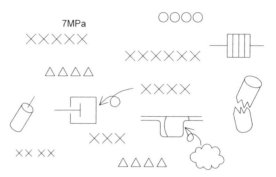

Figure 2.10   "Plane of Thoughts" where we record random thoughts which enter the mind

(b) Relation of thoughts diagram (Figure 2.11): Diagram that shown the relations among concepts that are scattered on the paper (also called a "chain of thoughts" diagram). We think about the relations among the concepts and collect those that belong to the same category, plus the relations with spatial curves, to express relations among the concepts. Putting numbers, following the sequence they enter the mind, on these curves (also called "links"), and words (also called "nodes") is sometimes useful.

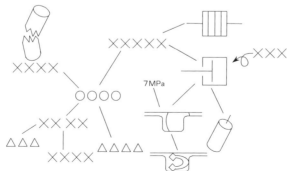

Figure 2.11   "Relation of thoughts diagram" for expressing relations and sequences of concepts that enter the mind

(c) Expansion of thoughts diagram (Figure 2.12): The basic flow of design proceeds Function → Mechanism → Structure, and this diagram shows this process. We record ideas that we selected as well as those we did not. We analyze and decompose the first given functional requirement into functions and functional elements. We think about a number of mechanism elements to meet each functional requirement, then select and decide which we think is the best. Deciding the mechanism element for each functional element is called "mapping." We then expand the functional element we selected and decided, add attributes like size or material, to form structures, and finally integrate the structures into the total structure.

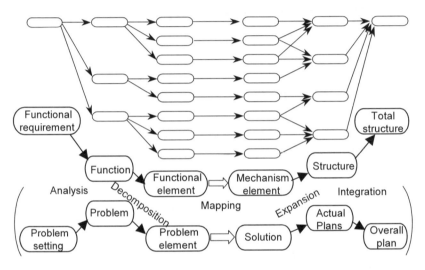

Figure 2.12   "Expansion of thoughts diagram" that expresses the design progress of Function → Mechanism → Structure

(d) The expansion of thoughts diagram is not limited to design processes, and is very useful when applied to the process of decision-making for solving problems. In this case, the thoughts expand in the sequence of Problem Setting → Problem → Problem Element → Solution → Actual Plans → Overall Plan.

(e) Selection and decision Diagram (Figure 2.13): Diagram that shows the process of selection. Recording the "selection and decision diagram" in this manner makes it easy for a reader to understand the decisions. The reader clearly visualizes which options the writer thought about and what judgment he made with which constraint. The constraints shown here (identified by double line frames with rounded corners) are usually not known in the beginning and they tend to clarify as selections and decisions take place.

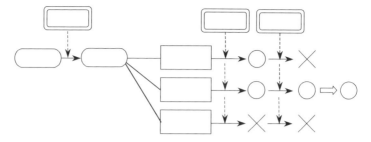

Figure 2.13 "Selection and decision diagram" for expressing the selections made while meeting the constraints

(f) Spiral of thoughts diagram (Figure 2.14): Diagram that shows how what was vague at the beginning gradually takes shape in a spiral manner. Inserting constraints and comments as the thinking proceeds makes the diagram easier to understand.

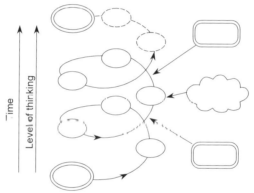

Figure 2.14 "Spiral of thought diagram" shows the mind process as it progresses with time

(g) Time history progress diagram (Figure 2.15): Diagram that shows the progress of elements, related constraints, and comments along the way. Assumptions implied on the first set problem but revealed later are written down in dotted line boxes to help clarify the process.

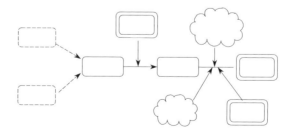

Figure 2.15 "Time history progress diagram" shows the progress

(h) Element relation spore diagram (Mandala diagram) (Figure 2.16): Diagram that shows the hierarchy among structural elements. Higher-level concepts are placed in the center, lower ones in the periphery, and the hierarchical relations among the structural elements are clarified.

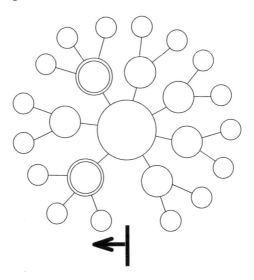

Figure 2.16 "Element relation spore diagram" shows structural elements and their hierarchy

(i) Structural element transition diagram (Figure 2.17): Diagram that shows the change of structural elements as the time progresses. The diagram proceeds from left to right with time and has comments and constraints added. Using double-lined boxes for governing elements within structures clarifies the transition.

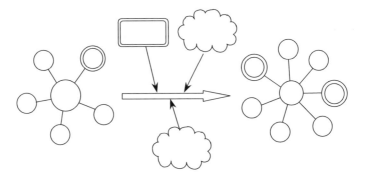

Figure 2.17 "Structural element transition diagram" shows the change of structural elements with the progress of time

(j) Structural element correspondence diagram (Figure 2.18): Diagram that shows how the structural elements correspond. Using double circles for primary elements clarifies the correspondence.

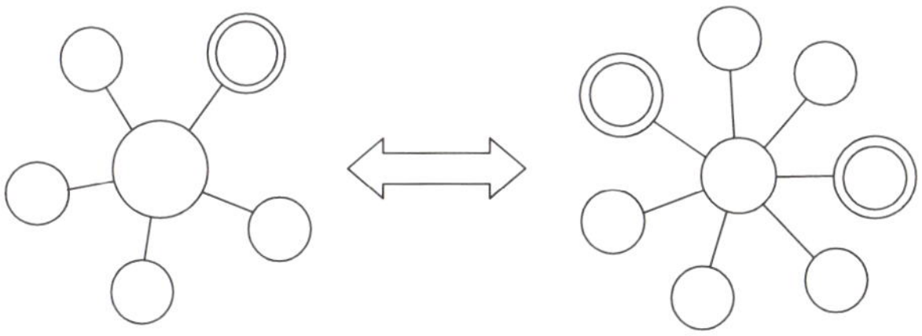

Figure 2.18 "Structural element correspondence diagram" shows the relations among elements

2.5 Example of Recording the Decision Process

Reading or viewing only the final decision itself does not allow us to understand the decision if the process of reaching it is not clear. We then cannot make use of the results from the decision. We want to know what happened in the course of reaching the decision, what the decision-maker thought, how he handled his hesitations, and how he reached the final decision. This section looks at a real experience by some of the authors of this book with a mountain hiking boot. It broke while we were climbing, and this example describes the decisions we had to make, and how to record the events so we can transfer the mind process of making decisions to the reader.

Title: "Our Temporary Fix of Tying up Mountain Hiking Boots that Broke While Hiking – How We Guessed the Mechanism of Destruction through Observation and Devised a Proper Countermeasure"

2.5.1 Describing What Happened

We first describe what happened, what we thought, what actions and decisions we took. We kept the descriptions brief by selecting only the necessary elements. When writing this part, we aimed at providing the substance of the decision and its overall structure, so that those wanting to learn about the decision can judge whether the subject is one for further study.

(1) **Event:** The rubber soles of mountain hiking boots came off while we were climbing a mountain. They broke near the glueline and not exactly on it. We guessed the mechanism of destruction by observation. We then collected strings from the party, tied up the boots to counter the forces on them, and recovered the original function of the boots. The temporary fix worked. We learned much from the actions we took.

(2) **Background:** We often went in for serious mountain climbing when we were younger, with tents and over several days or a week. We have the desire to climb mountains just as frequently as we used to, and so, instead of serious climbing once a year, we decided to go more often but in a casual manner so we in the middle age could hack it.

(3) **Course:** One day in the fall of 2001, seven graduates from the same lab (all authors of this book), took the mountain route from Tengendai in Yamagata prefecture to Mount West Azuma. As we walked the track, one of the party broke his boots (Figure 2.19). Their soles started to peel from the heel (A in Figure 2.19), and the crack gradually advanced forward.

Figure 2.19   The sole of the boot started to peel off

(4) **Countermeasure:** At first we thought we simply had to tie the upper to the sole. We did not have any spare bootlaces and first tried vinyl bags we were carrying but they quickly proved useless. We then started to pull strings out from our backpacks and jackets to tie up the boots. After much trial and error, we successfully tied up the boots to recover their original function (Figure 2.20, Figure 2.21). Without having to change our plans, we enjoyed our mountain climbing.

Figure 2.20   How we tied up a mountain hiking boot

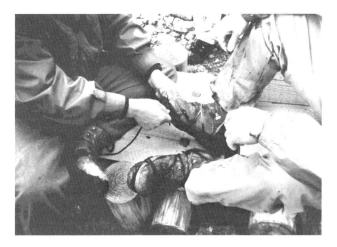

Figure 2.21   Tying up Mr F's broken boots

(5) **Mind process:** We repeatedly pondered over what happened. Our thoughts went into the micro-mechanism of breaking, planning a countermeasure, and executing the plan. In the regular course of design, the mind process proceeds Function → Mechanism → Structure; however, in this case, it went Observation → Guess the Mechanism → Countermeasure. We then tried and went back to the spiral of thoughts and onto devising a countermeasure (Figure 2.22). Figure 2.23 shows the mechanism of how the boots broke.

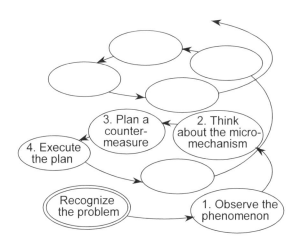

Figure 2.22   Spiral of thoughts from recognizing the problem to executing the solution

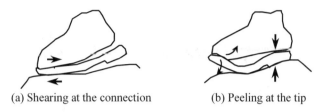

(a) Shearing at the connection   (b) Peeling at the tip

Figure 2.23   Mechanism of how the boot sole came off

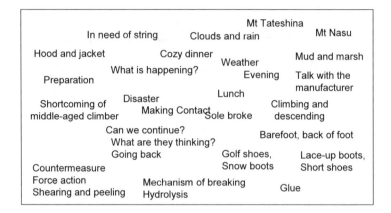

Figure 2.24   Thoughts that came to my mind as I saw the shoe sole peel (plane of thoughts diagram)

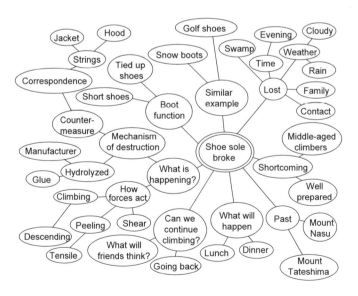

Figure 2.25   Relations among thoughts that came to my mind (relation of thoughts diagram)

As we saw the boot sole peel, many thoughts came to me. Things related to peeling, about continuing the climbing, countermeasures, and unrelated matters – all sorts of things just popped up in my mind. Figure 2.24 shows the plane of thoughts I constructed afterwards by remembering what came to my mind. Figure 2.25 is the relation of thoughts diagram that organizes the contents of Figure 2.24 and arranges the elements for easier understanding by the reader.

The peeling was not caused by a weakened gluing that attached the sole to the body. It was caused by repeated stress on the weakened material. I used to think boots had two functions: to cover the outside of our feet and the bottom of our feet. Figure2.26 shows the corresponding expansion of thoughts diagram. Further thinking and observation made me recognize the third function of "transferring the upper movement to the bottom". Figure 2.27 shows the relations among the functions of boots and their structure.

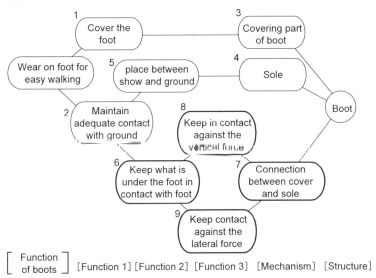

Figure 2.26 Expansion of thoughts diagram found by observing the boot sole peeling and expanding the boot functions and mechanisms

(6) **Knowledge:**

- Understanding the functional requirement and constraints through observation allows proper countermeasures even when in the mountain.
- Knowing the mechanism of deformation and destruction, and countering them led to the same conclusions with old-time technology (straw shoes and their tying up), which in turn have led to modern technology.

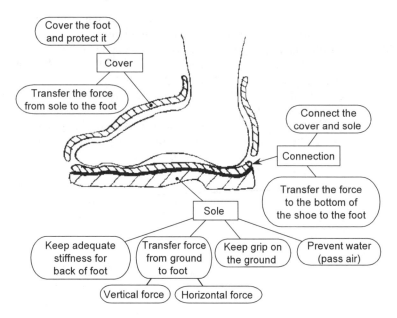

Figure 2.27   Structure of a shoe and function required to each part

(7) **Lessons:** Thorough preparation is necessary for climbing mountains. Maps, compass, flashlight, string, knife, matches, warm garments, and raingear are the minimum.

(8) **Sequel:** The story of these boots was the main topic for the party after we came down the mountain. The object that we finally produced (Figure 2.20) was very like the old straw shoe widely used in Japan. The functions required of what we wear on our feet are the same for the old straw shoes as for the mountain hiking boots. That is why the way strings were tied were in almost the same way once we were done.

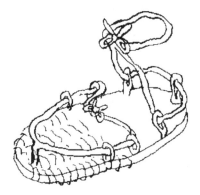

Figure 2.28   Structure of a straw shoe and the mechanism for wearing It

One of the authors (Hatamura) was evacuated from Tokyo to Tottori prefecture during World War II for safety. There he experienced the making of straw shoes. The sketch (Figure 2.28) is based on his memories.

Back in the lodge, the party warmed up their bodies and hearts with some help from alcohol and extended their talk to right-angle and parallel tying of bootlaces, people not tying bootlaces properly, people not knowing how to hold chopsticks, children not being disciplined, and so on. We just started with the mountain boot and enjoyed the evening talking about technology and culture.

2.5.2 Processes and Contents from This Event

For the reader who judges that this decision is something to learn from, we shall write down in detail the contents and processes of making the decision, the other matters related to the decision, so as to transfer the decision in a three-dimensional manner to the reader.

(1) When and where the event took place
     Date: October 6, 2001
     Place: We climbed Mount West Azuma, which is located north of Mount Bandai, which is north of Lake Inawashiro. Our climbing had a typical middle-aged schedule with a walk up a height of about 250 meters.
     Weather: Cloudy

(2) Detailed course of events: As soon as we started climbing, the boot soles of our friend (Mr F) started to come off, and when we reached the top the heels had almost completely peeled off. It started with the right boot, but the left one came off at the same time and both eventually completely disintegrated. It is not easy to make two things break at the same time. All of us wondered what kind of manufacturing control the makers had. Mr F offered to turn back to avoid troubling the rest of us, but we thought that if we stuck together we could do better than letting him return alone, and we could think our way through. We told Mr F so, and got him thinking the same way, *i.e.* that he could walk the whole course.

The photograph shows how we temporarily tied up the boots with the peeled-off soles (Figure 2.21). The peeling started from the heel and gradually advanced forward. When we first thought about tying them up with something, we tried a vinyl bag, which did not work. We then decided that we should use some strings, but we could not find any. None of us had the standard mountain hiking equipment of spare laces, so we had to them from our jackets and backpacks.

(3)   Mechanism of peeling: Here is a record of the peeling mechanism we
reasoned by observing the problem (*cf.* Figure 2.23). We quickly realized
that it was the shearing force in the remaining contact area between the
sole and the boot body that caused the peeling, however that force was
just a part of the mechanism. Further observation revealed that the force at
the tip of the crack pulled the two apart. In all, the horizontal shear and the
force to pull the two pieces apart caused by the shoe bending were
responsible for the peeling.

(4)   Process of trial and error with the tying: This paragraph describes the
process of tying up the upper and sole of the boot until we successfully
integrated them into one piece (*cf.* Figure 2.20). Until we understood the
entire mechanism, the tying was just groundless. We first tied the heel of
the sole to the upper with string (a) by one loop, but it was not easy to
walk with. So, we fixed the center with string (b). We then thought we
could make it work better and hooked the heel to the original bootlace (d)
by looping string (c) around it. This made it much more comfortable.
Firmly tying up the ankle string (d) with (c) and then pulling string (c)
with string (e), the boot upper and sole were held firmly together and
stopped coming apart. This step was crucial. Mr F walked without
worrying about his boots at all for the last hour and a half. This last string
was the most important one.

(5)   Expanding the thoughts to general footwear: I then realized that the tying
was the same with the straw shoes I used to make.
Let me discuss some about straw shoes (*cf.* Figure 2.28). I hear there
are different types of straw shoes. I used to make them myself and the
bottom piece is made of straws wound around four straight straws in the
long direction. One of the straw shoe strings starts from between the big
toe and the second toe, then splits into two parts which pass through rings
on both sides of the bottom piece and are tied together above the foot.
This string arrangement is the same with beach sandals, and I have shown
the arrangement in a Y-shape in the figure. The strings come horizontally
from the branch between the toes and turn upwards at around the middle
of the foot, then are tied on top of the foot to pull the bottom piece against
the foot to keep it with the foot. The heel part is pulled up towards the
ankle. A straw shoe attaches itself to a foot at three locations: between the
toes, on top, and at the ankle.

(6)   Generalizing knowledge and structure: Looking at the stringing alone, we
realized that how we tied up Mr F's boot (Figure 2.20) resembles a straw
shoe (Figure 2.28). This means that the stringing system has to be this
way to accomplish the function of keeping our foot and what is
underneath it (a sole or a straw shoe) together when we walk. We realized
that the straw shoe has a logical structure.

(7) Relation of concepts and example: When we trace the process of the peeling of the mountain hiking boot sole and how we countered the problem, we can draw the spiral diagram of thinking about the micro-mechanism of how the boot sole disintegrated, planning the countermeasure, executing the plan, observing again, iterating the plan and executing it (Figure 2.22).

(8) Expression with an expansion of thoughts diagram: We can express the process starting with identifying the function of the boot and eventually reaching the tying system in an expansion of thoughts diagram (Figure 2.26). Among the functions of the boot, those of "protecting the foot" and "protecting the bottom of the foot" prevailed even when the sole came off; however, we then recognized the important function of "keep the foot and sole in contact to transfer force," which did not.

  The actual thinking progressed by going back and forth rather than in a steady left to right movement of Function → Mechanism → Structure.

(9) Generalizing the function and mechanism relation: Our further thinking revealed that what came out as the result was almost the same as what had been in existence for a long time. In other words, pursuing the same functions led to the same basic structure. We can say that there is an inevitable relation beyond time between a function and the mechanism to realize it. There is an intriguing universal generalization in the world of technology.

  The tying reminded me of ancient Roman warriors in movies wearing sandals tied like the straw shoes. Some of the current running shoes have their markings modeled after those Roman sandals.

(10) Finding similar events: When we arrived at our lodge that evening, we took a closer look and found the stacked layers of rubber between the boot sole and the boot body had turned into crumbs of a layered cake.

  I had expreienced similar failures before. Once my old golf shoes on one of the sole suddenly came off. What broke was the material near the glueline, but not the glueline itself. I first thought it was the glueline and complained to the manufacturer, who replied "Such peeling is common in shoes that are not worn often and kept in storage." They said the moisture inside hydrolyzes the material producing H or OH, and that reaction causes the internal destruction.

  Another example is snow boots. I was on a Shinetsu line train (the Nagano bullet train construction shut down the local tracks near Usui summit and only part of Shinetsu line is still in service today) when my snow boots suddenly started to disintegrate, and by the time we reached Nagano station I was barefoot. I had to buy rubber rain boots at a shoe store near Nagano station. I wished the boots had broken into something like beach sandals with soles. Unfortunately, the bottom of a snow boot is in one piece with the rest. So my snow boots had something left on the top of the foot but the bottoms of my feet were in direct contact with the road. I walked through the town of Nagano in those boots to the shoe store. The people at the shoe store laughed but told me it was a common thing to happen.

Asking around, I found that there were quite a few people that had experienced similar troubles. The common factor among them was "occasional use" only a few times a year. Shoes worn daily do not have such troubles. Shoes that break abruptly are those kept in boxes for years or ones that are called for once or twice a year for middle-aged mountain hiking or middle-aged skiing. They break not at the glued surface but rather disintegrate like cake crumbs. Observing the failed material, we saw a ragged surface, probably due to hydrolyzed plastic. The manufacturer wrote to tell me that shoes break in a such manner. I was somewhat upset by such a comment, but having now heard more of such happenings, I think that it is something we have to put up with.

> There have been accidents caused by ski boots breaking during skiing. I have also heard about a recall triggered by the ski boots breaking in the middle of competition. Such incidents were probably due to the same mechanism as described here.
>
> If this happened on a mountain 3,000m high, there would not be much we could do about it. It is horrifying to think of getting lost in the mountains because the soles of your boots broke off.

(11) Understanding the chemistry of the destruction: It is a commonplace among chemists that materials made from polymerization break easily when hydrolyzed. If we think that way, this trouble with the shoe sole is better described as "destruction", rather than "peeling".

(12) Model to build in the mind: When we look microscopically at the shoe sole near the glued surface, the material has many small holes, *i.e.*, a number of microscopic pumps are connected to one another. In other words, we will imagine a porous pump model. If we use the shoes daily, the water is pumped out, but if we store the shoes for an extended period of time, the pumps are kept still, causing hydrolyzing in the material (Figure 2.29).

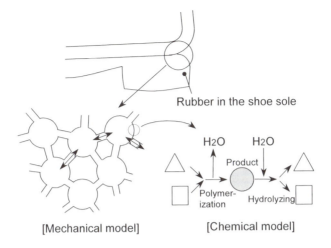

[Mechanical model]   [Chemical model]

Figure 2.29   Model to build in the mind about the shoe sole peeling

Manuals Stop Your Thinking Process

Manuals are all over the place. When a scandal is revealed, a typical countermeasure is "We will strengthen our management" or "We will revise our manual." Can such actions really prevent scandals? I always wonder about it. Now I have started to think that these problems are caused by the laziness of large corporations that force manuals on their employees who eventually end up working without thinking.

The production area cannot run proper lines without the manual (documented job standards). Manuals are absolutely necessary. So, why did I end up thinking that manuals are the source of problems? I realized that manuals themselves do not cause problems but rather the mind or the thoughts of the humans that use the manual. That's it! It is the dead manual! The thought that "I am following the manual so there is nothing wrong that I am doing" is the wrong course of thinking.

I read, the other day, the book *Economy of Yoshinoya* coauthored by the president of Yoshinoya, Mr Shuji Abe, and Professor Motoshige Itoh of the University of Tokyo. President Abe, in the book, says "Manuals are there to follow, but also to change." I believe this is the true meaning of a manual.

There are plenty of examples of dead manuals. Whenever I enter a store and hear the flawless but also emotionless greeting, I believe the recession in Japan will still continue and unbelievable failures will keep mounting up. We must not blame it on manuals. We are the ones to be blamed who stopped thinking about each step just because following the documents is so easy. Let's start correcting our actions.

3

Decisions in Design

3.1 Questions about Design

Chapter 1 discussed the mind process of how we think, wonder, and decide when we solve problems. Design is one such problem. Depending on the level of the designer with his mind deep in the process of designing, a number of questions arise including the following:

- Does designing mean just following samples that are already available?
- Can we follow manuals (written procedures) to design? Or if we carefully search standards, can we find the answer?
- Where do we start when we design?
- What is the procedure for making a design?
- How do we set the design goal? What are the free-to-set variables? Dependent variables? Constraints?
- (When we look at a drawing,) how was each item designed?
- (When we look at a product based on a design,) how were its shape and structure determined?

Questions come up one after another. Many of these relate directly to technology or science but most of them are questions about the act of design itself. When a student mimics design, he has a "sample" to follow. Modifying an existing design starts from a "former design." A design with an established method proceeds following a "design manual (procedure)." In these cases, it is easy to know where to start and how to complete the design. In creating something new, however, there is no sample, no former design, no manual. The designer has to produce everything for himself. In this situation, the most important thing to know is how decisions are made in design.

Those who watch or read actual products, design drawings, or design records – we will call them followers – want to understand the process of making decisions or selections the original designer made from these data. The follower's mind tries to track the designer's mind process in the reverse direction, as Figure 3.1(a) shows. If we call the original design process the "forward operation," then what the follower does is the "reverse operation." If we start from the results after a structure is decided, the designer will have recorded the outcome of selections that are usually isolated and simplified, and the path to reaching the decision will be hard to follow. The follower, given the simplified fragments, cannot understand the decision because they do not fit his mind process. If someone cannot understand it, then he cannot use it. The follower, therefore, infers and searches the path starting from what results are available until he thinks he has understood the process. The path found in this manner just traces the designer's decision process of the mind in the reverse direction.

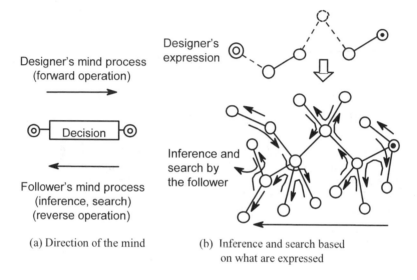

(a) Direction of the mind

(b) Inference and search based on what are expressed

Figure 3.1   Followers want to know the decision processes

3.2   The Mind's Decision Process During Design

Figure 3.2 shows how the decisions proceed when we design something totally new (we can call this process "creative design"); *i.e.* Functional Requirement → Function → Mechanism → Structure). If we put the process in words, analyzing and decomposing the functional requirement leads to a set of functions. The original functional requirement and decomposed functions are abstract concepts. We then map each function to the physical domain of a mechanism to realize the function. Then we turn the mechanisms into a final structure by adding attributes like size or material and synthesizing them. Figure 3.3(a) shows further details of this mind process in design. Our thoughts expand during the entire process and even within the vertical analysis and decomposition of functions from high level to low level, or in the horizontal process of mapping, expansion, or synthesis, we hesitate in making selections at each stage (Figure 3.3(b)).

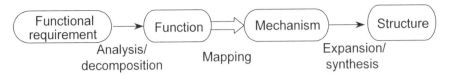

Figure 3.2 Basic mind process in design

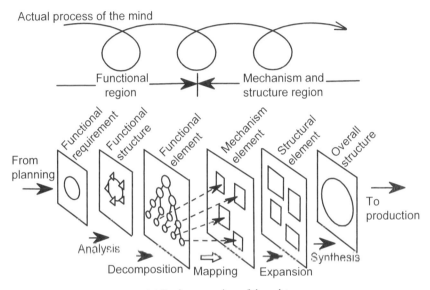

(a) Basic expansion of thoughts

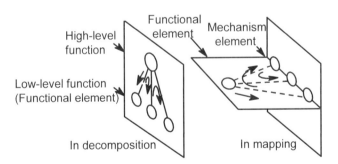

(b) Hesitation and selection at each stage

Figure 3.3 Basic expansion of thoughts and making selections in creative design

The process takes a rough path of Function → Mechanism → Structure when we make decisions, but the actual decision process of the mind does not proceed so simply in one shot: evaluation and feedback take place at each stage. As a result, the process loops through function, mechanism, and structure. The design repeats this cycle as it proceeds. Figure 3.4(a) shows this spiral process. Furthermore, this spiral curve itself passes search, hesitation, evaluation, and selection. A closer look at the spiral curve thus reveals a smaller spiral and the entire mind process shows a "double spiral of the mind." Note that the "double spiral" here is different from the same term used for DNA structures.

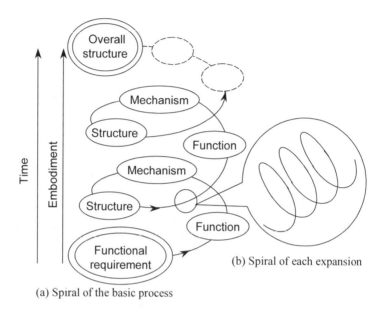

(a) Spiral of the basic process

Figure 3.4   Double spiral of the mind process in design

Let's look at how the mind process corresponds to the actual design process (Figure 3.5). The spiral in Figure 3.4 appears, when viewed sideways, as the zigzag course in Figure 3.5(a). On the other hand, the actual design starts by receiving the primary specification (that lists the functional requirement and constraints) from the planning group then proceeds Sketch → Schematic → Final Plan Drawing. Both the sketching phase and schematic phase repeat many times as the mind deepens its understanding. The final plan drawing lists all the information related to the product to build. The actual design process is thus tightly coupled with how the human mind develops its thinking.

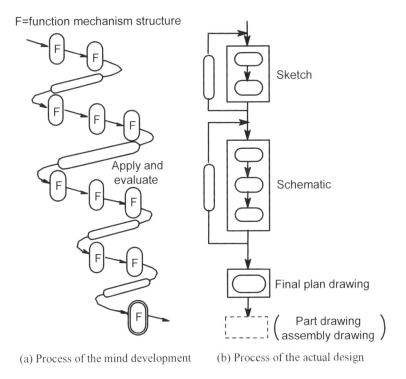

F=function mechanism structure

Apply and evaluate

Sketch

Schematic

Final plan drawing

Part drawing
assembly drawing

(a) Process of the mind development   (b) Process of the actual design

Figure 3.5   Relation between the mind process during design and the actual steps of design

3.3   Constraints and Evaluation of Design

Figure 3.6 shows the process of making a design decision. We start from a new project, make a tentative decision based on the given constraints, repeatedly feed back the results for evaluation, and eventually make the final results. This subsection discusses the constraints and evaluation that relate to decisions in design.

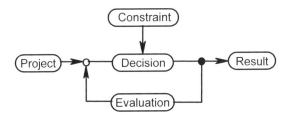

Constraint

Project → Decision → Result

Evaluation

Figure 3.6   The relation between the mind process during design and the actual steps of design

First, the constraints. There are many constraints, as Figure 3.7 shows, on designs. The figure shows these constraints in the order that they come to the designer's mind. Purely technical constraints are on the left side, starting from function/mechanism in the lower left-hand corner. As the design progresses, those on the right start to weigh more. The most significant ones for the overall design project are cost, turnaround, and method of manufacture. Figure 3.7 shows the constraints about design in the narrow sense. As technology and industry advance, and the results of the design affect our lives or society, constraints shift from the early technical or economic limitations to social, environmental, or ethical constraints, as Figure 3.8 shows. The designer must bear this deeply in mind, or social aspects of the design will even result in the evaluation that "Designing is evil."

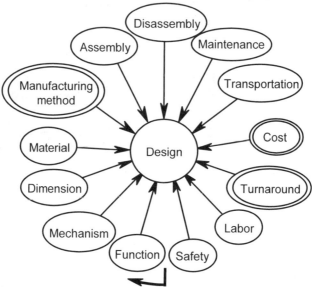

Figure 3.7   Constraints associated with design

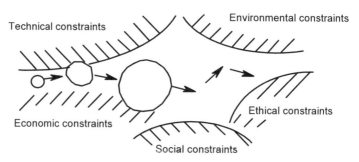

Figure 3.8   Transition of constraints with advancements in technology or the industry

Engineers that stay in one area have little interaction with the outside world and tend to become insensitive to changes there. Many of them continue in their way of doing things when the methods may no longer be acceptable to society. Such inflexibility often results in a failure or an accident. The actions the engineers take upon such an event occurring can determine the fate of corporations. We shall thoroughly study the Monjyu accident at the late Power Reactor and Nuclear Fuel Development Corporation.

Next, let's look at evaluations made in decision processes. For the design decision in Figure 3.6, the feedback evaluations include established evaluations, and often the individual taste of the designer like experience, or knowledge, as we discussed in Chapter 1. For the design process in Figure 3.5, the evaluations repeated during the sketch or schematic are mixtures of these evaluations.

When the design contents take shape, evaluations through feedback from the market are organized much better than those that go on inside our head and are easy to understand when viewed from outside. The process of manufacturing generally proceeds as Figure 3.9 shows, in the sequence Planning → Design → Manufacture → Sales → Use. In addition to evaluation and its feedback, other feedbacks for evaluation, including Manufacture → Design, Sales → Design, Use → Sales, Sales → Planning, and Design → Planning, are also important. Among these, the most important for production are the two evaluations of Design → Planning, and Sales → Planning.

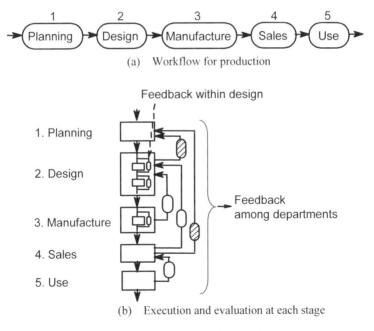

(a) Workflow for production

(b) Execution and evaluation at each stage

Figure 3.9 Constraints associated with design

3.4   Describing the Design Content

At the beginning of this chapter, in Section 3.1, we described how to display the mind process of making design decisions in a form that others can understand. This is important in supplying answers to the many questions that arise in design. This fact is also true as regards accumulating, conveying, and learning technology.

Figure 3.10 shows the documents related to design. Single-line circles indicate documents for regular design, double-line circles are those that are highly important, and broken-line circles are articles usually skipped but should in fact be recorded. Articles include plans, proposals, design records, fabrication procedures, assembly procedures, service procedures, operation manuals, product datasheets, patent documents, design manuals, design evaluations, engineering reports, trouble reports, and so on. Articles that should be fed back to the designer include trouble report, product evaluation, market survey, and product planning evaluation.

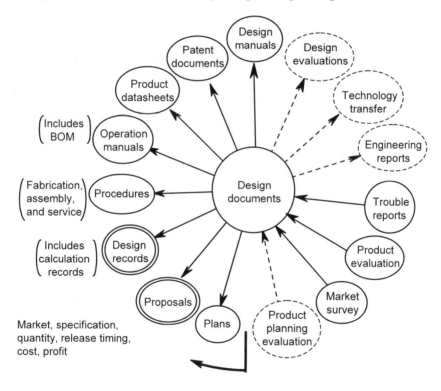

Figure 3.10   Design documents

The most important two among these are the proposal that discusses the market, specification, quantity, timing, cost, and profit, and the design record that describes what the designer thought and how he made decisions. Without these two, no matter how perfect the drawings (part drawing, assembly drawing, and BOM) may be, we cannot transfer the technology itself to others. In others words, others may be able to build the product but there is no further development and the technology just dies without recovery.

Other documents that are not usually produced that but the designer should make known include the design evaluation report, technology transfer document, and trouble case reports. For the transfer of technology in Chapter 2, describing the evaluation of a plan is the most important information source available to the designer from the outside and should really be produced.

3.5 Designers Want to Know the Reason for the Designs

Part drawings and assembly drawings merely describe what to produce. The schematic lists all information about what to manufacture; however, it does not record why the designer reached the conclusions. The follower designers want to know not just the decisions themselves, but also how they were reached. The drawings or design records, however, lack the descriptions of scenarios, *i.e.*, what other options the designer thought about, what he tried, and what mistakes he had made in the past (Figure 3.11).

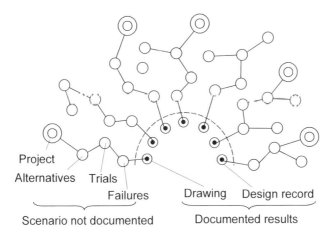

Figure 3.11 Drawings nor design records do not record the scenario of reaching decisions

For example, let's look at the reason for setting dimensions, one of the scenarios a follower designer wants know (Figure 3.12). Some are determined by functional requirements or strength requirements, some are set by outside standards, some by the interface with other parts, and some by manufacturing constraints or the shape of the machining tool. Normally these reasons are not explicitly recorded, but such dimensions have their own reasons and follow logic. Such items of background information, however, are not relayed to the follower designer. Other dimensions may be set for reasons such as, a round number, instinct, hunch, following traditions, converting English units to metric, or following existing products that had no logical reason either. These examples, however, are not as bad as those that have unknown reasons not recorded anywhere. Changing dimensions that have unknown background reasons can cause trouble and these are indeed the ones that follower designers have most trouble with.

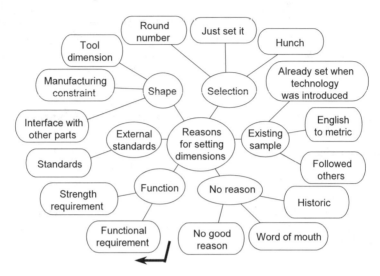

Figure 3.12   Examples of unexpressed reasons for setting dimensions

Figure 3.13 shows the selection and decision-making processes when the designer has several candidates. This example of straight motion has the candidates of bellows and linear bearings, a hydraulic cylinder, a diaphragm, or a stacked piezoelectric device. The designer evaluates each candidate for its stroke, resolution, and how easy the driver is to handle, and the sample shows the case where a hydraulic cylinder is selected.

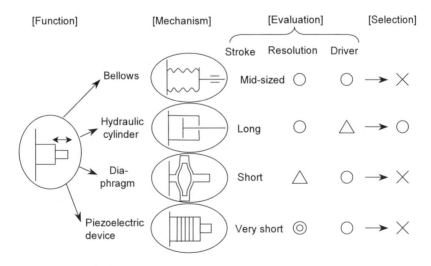

Figure 3.13   Examples of unexpressed reasons of setting dimensions

The designer then wants to know what happens when he selects each option. He would really like to know how a similar design performed in the past. Officially recognized drawings like the part drawings or assembly drawings are sometimes called "explicit drawings" to distinguish them from "hidden drawings" that are not officially recognized but often drawn by designers with goodwill (Figure 3.14). Among those hidden drawings, the results diagram that records how the product performed should always be inserted the design records. Although most designers do not produce them, the one most wanted by follower designers is the "options diagram" that records what other alternatives the designer wondered about and the "decision diagram" that explains what the designer thought and how he made decisions. The true transfer of design knowledge takes place only if the designer produces these diagrams.

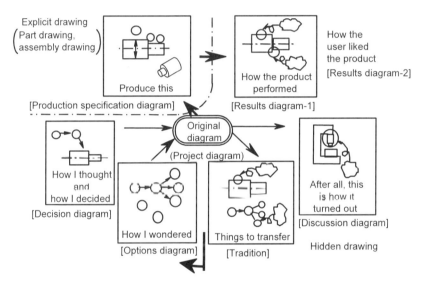

Figure 3.14   Explicit drawings and hidden drawings

3.6 Design Support Systems the Designer Wants

The designer wants assistance at each stage of the design (Figure 3.15). In the sketching stage, the follower designer looks for information about what earlier designers thought and wondered; in the iterating schematic stage, what references they used; and in the final schematic stage, what alternatives were available and what made them make the selections and decisions. These constitute information about the mind activities. The designer also wants to know what happened after production and how the products were used. If the situation allows, the follower designer wants to know, throughout the entire process from sketch to final schematic, what earlier designers thought and what they learned.

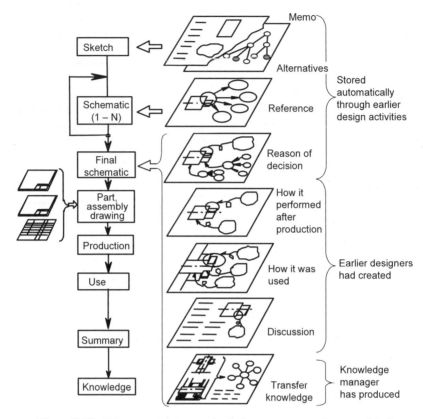

Figure 3.15   What external support the designer wants at each stage of design

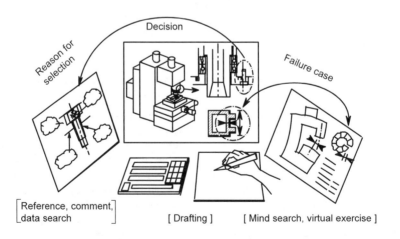

Figure 3.16   CAD system the designer wants

Figure 3.16 shows a concept of a system that, by meeting all the above requirements, aids the designer's mind activities. Figure 3.16 shows the main screen in the front with the design overview and part details; when the designer clicks an object in the screen, the reason for selection or troubles related to the object appears. When he picks a selection displayed in the countermeasure to the trouble, then a parameter or other design attributes are automatically selected in the main screen. Another screen serves to support the mind activities as search, and CAE or past data help the designer make decisions about design parameters to approach the target design.

What the designer wants is a hierarchical display of data (Figure 3.17). Hand-drawn schematics carry every piece of information and if all those are placed on a single screen, the designer will be lost not knowing where to start looking, and will be forced to arrange the information in a hierarchy within his mind. So he will fail to build the information he wants within his mind. CAD data already in the system must be arbitrarily selectable from the outside to show, *e.g.*, just the shape and dimensions or shape and material.

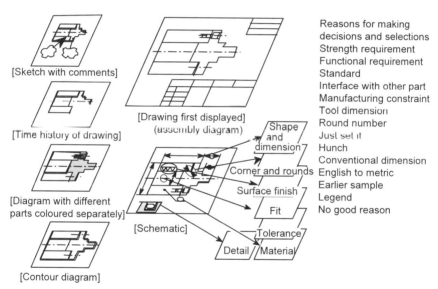

Figure 3.17   What diagrams and information a CAD system should display to the designer

In addition to these hierarchical displays, the designer also wants the following: a drawing with comments on each shape and description, a drawing that highlights where the same comment appears, a drawing that shows the process advancement with time, and a drawing that shows different functions with the same family of colours.

The designer absolutely needs to know how the thoughts expanded and how each selection was made. Figure 3.18 shows the expansion of thoughts diagram for a fine-positioning machine in vacuum. It analyzes the functional requirement of "fine positioning in a vacuum" and splits it into linear guiding, motion, and detecting position. The diagram then further decomposes these into unit functions and selects a mechanism for each unit function. It then evaluates a number of mechanism candidates under newly recognized constraints and decides to either adopt or discard each one. In this case, the final structure conducts rough positioning with a straight guide with vacuum grease, a ball screw, a stepping motor, and an optical linear scale, then fine-tunes the position with a parallel plates structure, a stacked piezoelectric device, and detects position with strain gauges, all put together in series. This expansion of thoughts diagram has a number on each entry so that the reader can later track the sequence of thoughts.

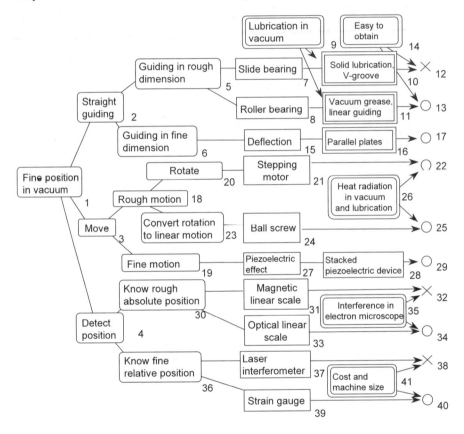

Figure 3.18   Expansion of thoughts diagram for fine positioning under vacuum

As we explained in Section 3.5, the designer wants information about former designs and products. Figure 3.14 showed hidden drawings that show such information. The designer wants to have immediate access to such information in the hidden drawing by linking the hidden information with 3D CAD data. Figure 3.19 shows an example with a hydraulic cylinder that links 3D CAD and hidden drawing information. The figure shows alternatives, reason for decision, results of production, performance, discussion, and knowledge transfer in the form of comments about the piston rod material, method of fabrication, and surface finish, in addition to the regular drawing information. Furthermore, the words and phrases in the comments go through natural language processing so that similar concepts, their relations, past failures, and troubles can be searched.

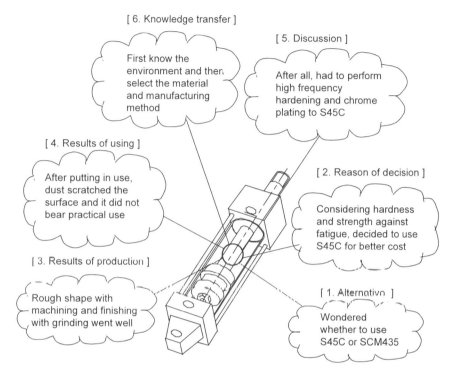

[6. Knowledge transfer]

First know the environment and then select the material and manufacturing method

[5. Discussion]

After all, had to perform high frequency hardening and chrome plating to S45C

[4. Results of using]

After putting in use, dust scratched the surface and it did not bear practical use

[2. Reason of decision]

Considering hardness and strength against fatigue, decided to use S45C for better cost

[3. Results of production]

Rough shape with machining and finishing with grinding went well

[1. Alternative]

Wondered whether to use S45C or SCM435

Figure 3.19 Hidden drawing comments shown to the designer – Hydraulic cylinder –

To meet the requirements of Section 3.5, we planned a system in this section, and actually built the system under the sponsorship of the Ministry of International Trade and Industry. The project took us three years to finish. See Section 4.7; "Built 'Creative Design Engine' for Assisting Idea Generation."

4

Sample Decisions in Design

The chapters so far have discussed what decisions are, how to describe them, the need for transferring the decision process in design, and what to transfer. This chapter shows how the design decision processes went with the following design samples.

(1) Hydraulic cylinder: We start from one of the most common machine elements, a regular hydraulic cylinder, and show how the designer decides its structure. We then go up to a higher level of concept and review the design.

(2) Torque sensor: One of the basic elements for an intelligent machine is the force sensor. This example starts from the functional requirement for setting the mechanism and structure. It shows the effect of computer support by using a 3D CAD tool by showing how deformation and stress change and the designer changes the design parameters.

(3) Positioning table: This example explains the positioning device which is the most fundamental element for machine tools, measurement machines and test machines.

(4) Intelligent grinding tool for producing flat wafer surfaces: The grinding force reaction tilts the grinding wheel and interferes with producing a perfectly flat surface on large wafers. This sample shows a grinding tool that tilts its spindle axis and table, in response to the grinding force, to finish a perfectly flat surface on a wafer. This subsection shows the mind process in building this system.

(5) Nanomanufacturing World: Built a system that manipulates a robot within a vacuum chamber and allows watching the nano-operations via an electron microscope. The subsection reviews the mind process in planning and obtaining funding for the project.

(6) Control system for automatic grinding of turbine blades: Designed and prototyped a grinding system for automatically producing turbine blades with curved surfaces. This subsection explains the design of the adaptive control system.

(7) Creative design engine (software) for assisting idea generation: Here we review how we constructed a design support system for designing

something totally new with no prior samples to rely on. It took us three years to actually build this software.

(8) Stirling engines: Regular gasoline engines are internal-combustion engines and Stirling engines are external combustion engines. Starting from its basic cycle, the students think of a variety of different Stirling engines, build and test them. The examples in this subsection were actually built by juniors in the mechanical engineering department at the University of Tokyo.

4.1 Hydraulic Cylinder – the Mind Process of Designing One

A hydraulic cylinder is a machine element that supplies pressurized oil to a cylinder, generating force on the piston to produce straight movement. Many machines use them as actuators. In practical machine design, the designer does not carry out detail design; instead he sets the operational pressure and stroke, then selects a model that meets the constraints from a catalog. Its mechanisms and structure have elements common to all machines. Reviewing the mind process in designing one in detail is quite effective in describing the mind process of creative machine design and so we picked it as the first example here.

A design starts by setting the design specification based on the project requirements. For a hydraulic cylinder design, as Figure 4.1 shows, the project starts with a plan to build a hydraulic cylinder with a certain stroke and force and sets a number of design specifications under the constraints of allowable piston size or operational pressure. Once these specifications are set, other constraints like rough work space volume and connection interface are set, and then following the explanation in Chapter 3, the mind process of design expands along the line of Function → Mechanism → Structure.

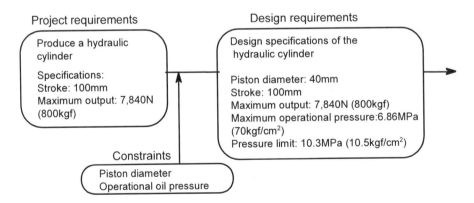

Figure 4.1   Project and design requirements for designing a hydraulic cylinder

Figure 4.2 shows the base process of expanding the thoughts in designing a hydraulic cylinder. First in the functional domain, analysis and decomposition take place (Functional requirement → Functional structures → Functional elements); then each functional element maps to a mechanism element then in the mechanism and structure domains, expansion and synthesis occur (Functional element → Structural elements → Overall structure). During the actual design, the processes do not proceed in a single direction, and thoughts go back and forth between each step and the next to determine the design and eventually reach the overall structure.

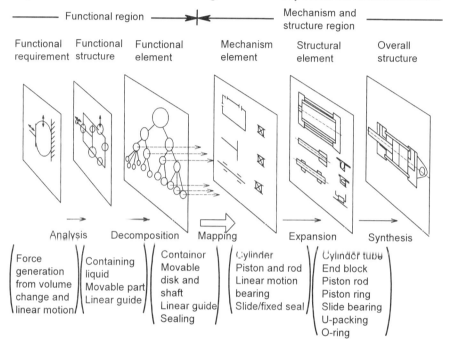

Figure 4.2   Basic expansion of thoughts for designing a hydraulic cylinder

In real design, after expanding (Function → Mechanism → Structure), the process goes through some adaptation for discussion and evaluation, then repeats (Function → Mechanism → Structure). The whole process repeats this iteration.

A regular hydraulic cylinder design has a piston inside a cylinder, and pressurized oil enters and exits the cylinder to move the piston back and forward. The structure is already set and without going all the way to the function and mechanism, the designer picks the dimensions and material. In this subsection, we will trace back the (Function → Mechanism → Structure) expansion and review what functions are required to hydraulic cylinders and what other options are available to meet those functions. We recognize from this review that the original functional requirement of "Force generation from volume change and linear motion" undergoes expansion and evaluation to reach the final overall structure. We will examine the hydraulic cylinder based on these functional requirements and then discuss the process of thought expansion if we were to design one from scratch. Figure 4.3 shows the entire expansion of thoughts process from the original functional requirement. The following discussion refers to this figure.

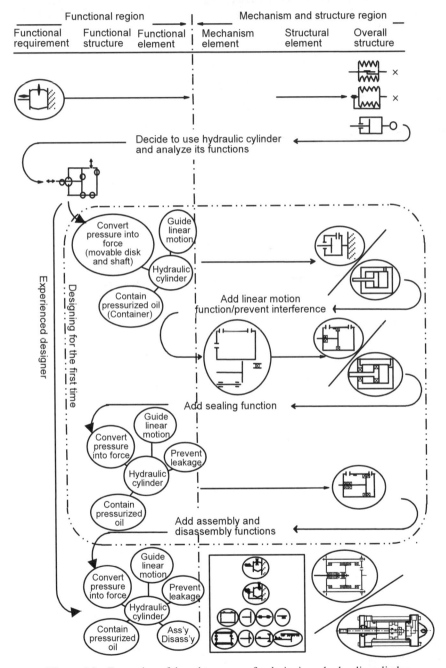

Figure 4.3   Expansion of thoughts process for designing a hydraulic cylinder

First, Figure 4.4 shows the functional requirement of "force generation from volume change and linear motion" mapping into the first-level mechanism. Methods of volume change include elastic deformation of rubber bags or bellows, and container and movable disk. Further adding the mechanism for linear motion produces a variety of options in Figure 4.4. The functional requirement calls for a machine element of a hydraulic cylinder and we select the combination of cylinder and piston in (B). This selection, however, is not always the one to choose; for example, if the requirement specifies a 1mm stroke with minimum stick slip, then we should select the combination of bellows and linear guide mechanism.

Figure 4.5 shows the function required for each component of the cylinder and piston assembly. The figure shows that the assembly requires a number of functions. Maybe we did not recognize them when we previously designed a hydraulic cylinder, but each of them is important and missing even one of them can cost the existence of the hydraulic cylinder itself. Setting the mechanism clarifies functional requirements and constraints that we did not know previously.

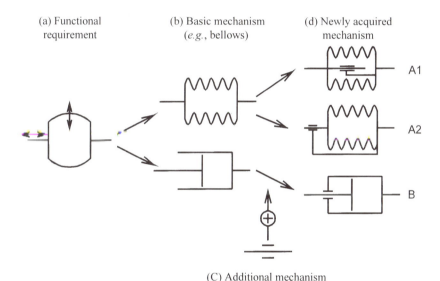

Figure 4.4   Process of determining the mechanism for designing a hydraulic cylinder

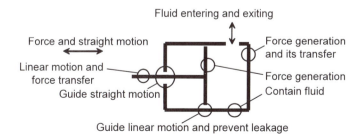

Figure 4.5   Structure of functions required to a hydraulic cylinder

Figure 4.6 summarizes the functions: (i) as a container for holding the pressurized oil, (ii) as a movable disk and a shaft that transfers the force, (iii) as a linear guide that allows linear movement of the shaft, (iv) as leakage prevention, and (v) for assembly and disassembly. Further breakdown of these functions gives the details in the figure, shown with the names of the machine parts that realize these functions.

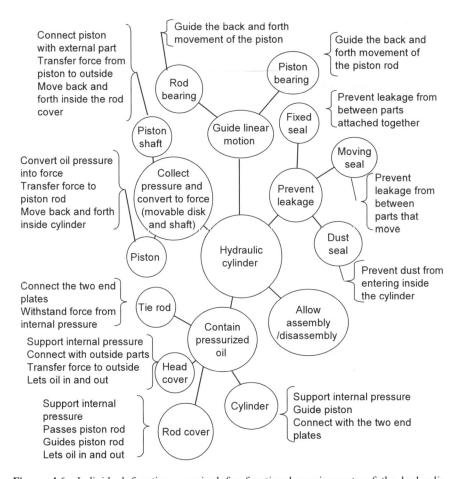

Figure 4.6 Individual functions required for functional requirements of the hydraulic cylinder

Figure 4.7 reorganizes the functional structures to show the structures in the functional domain. The bottom of the figure shows some machine parts that bundle some functions as before. Each of these is called a unit function. Each unit function in the functional domain "maps" to a unit mechanism in the mechanism domain, as the lower part of Figure 4.7 shows. The mapping of an abstract function to an actual mechanism is one of the most essential activities in design. A single unit function maps to one final mechanism; however, the number of mechanisms that the designer can list to meet a function strongly depends on the experience and the study by the designer. The functional domain and the mechanism domain correspond not only in the lower level unit function and unit mechanism but, as this diagram shows, also to the higher level bundles. This fact demonstrated the development of thoughts in the design of the structures of the mechanism. Figure 4.8 is the detail line diagram of the mechanism.

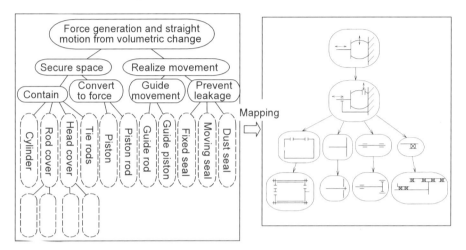

Figure 4.7 Correspondence between functional domain and mechanism domain for a hydraulic cylinder design

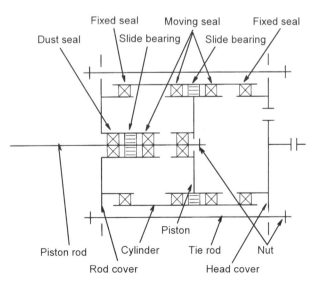

Figure 4.8.   Detail mechanism line diagram of a hydraulic cylinder

The discussion so far applies to experienced designers. If you are designing for the first time, things do not proceed so smoothly. The following paragraph simulates the first-time design.

In the physical domain (mechanism/structure domains), the designer adds dimensions to the mechanism (adding sizes, thicknesses and other dimensional attributes) and expands to a structure. The first task is to expand the container into a structure (Figure 4.9). Once the designer adds dimensions to the detail diagram of the mechanism (Figure 4.9(b)), he realizes the interference between the piston and the oil port. By moving the port to eliminate the interference, he then recognizes that he needs to seal between rod and container, and between piston and container.

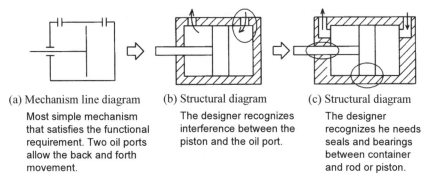

(a) Mechanism line diagram

Most simple mechanism that satisfies the functional requirement. Two oil ports allow the back and forth movement.

(b) Structural diagram

The designer recognizes interference between the piston and the oil port.

(c) Structural diagram

The designer recognizes he needs seals and bearings between container and rod or piston.

Figure 4.9   Expanding from mechanism to structure with the hydraulic cylinder design

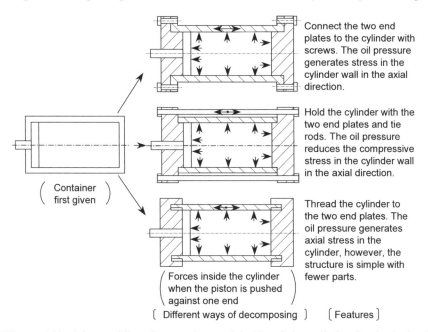

Connect the two end plates to the cylinder with screws. The oil pressure generates stress in the cylinder wall in the axial direction.

Hold the cylinder with the two end plates and tie rods. The oil pressure reduces the compressive stress in the cylinder wall in the axial direction.

Thread the cylinder to the two end plates. The oil pressure generates axial stress in the cylinder, however, the structure is simple with fewer parts.

(Container first given)

(Forces inside the cylinder when the piston is pushed against one end)

[Different ways of decomposing]   [Features]

Figure 4.10   Disassembling the container and inside of the cylinder for the hydraulic cylinder design

The structure in Figure 4.9 meets the required function; however, it clearly is impossible to produce: we cannot take the structure apart or build it. We will discuss how to take the cylinder apart as an example (Figure 4.10). We can think of various ways of closing the two ends of the cylinder with plugging plates and how to fix the plates. We will adopt the method of fixing the plates on the cylinder using long screws.

Once we set the structure, we draw the "force flow diagram" (Figure 4.11) to evaluate how the force flows inside the hydraulic cylinder parts during operation.

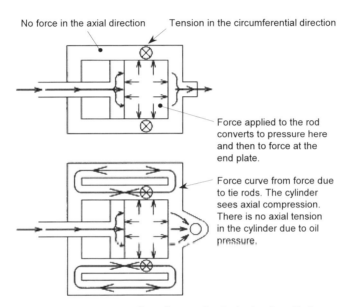

No force in the axial direction   Tension in the circumferential direction

Force applied to the rod converts to pressure here and then to force at the end plate.

Force curve from force due to tie rods. The cylinder sees axial compression. There is no axial tension in the cylinder due to oil pressure.

Figure 4.11   Force flow diagram for the hydraulic cylinder

Figure 4.12 shows the three types of sealing functions: a sliding seal between parts with relative motion, a fixed seal when there is no relative motion, and a dust seal to prevent dust from entering gaps between moving parts and binding them. Here we will use U-packing, O-rings, and dust seals to meet these functions.

In designing a hydraulic cylinder, the area where the piston connects to the piston rod has all the issues we need to think about, namely, part shapes, surface roughness, mating, assembly, and disassembly. Figure 4.13 shows what the designer thinks in designing this area. The figure shows that the designer has to think about a number of issues and runs virtual exercises to make decisions.

After the detail mind process of design so far, the designer obtains the final mechanism diagram shown in Figure 4.13, and the overall structure with dimensions shown in Figure 4.14. Designing even the simplest hydraulic cylinder follows the mind process of design in Chapter 2, and knowing this process allows a proper virtual exercise to come up with a better design.

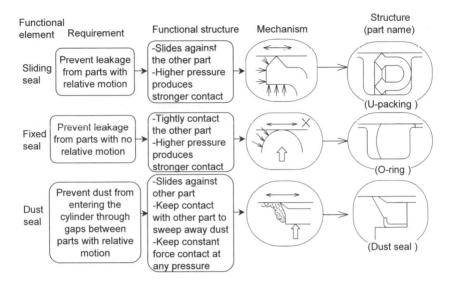

Figure 4.12 Expanding the required function of preventing leakage from the hydraulic cylinder into mechanisms and structure – selecting purchased parts for designing a hydraulic cylinder

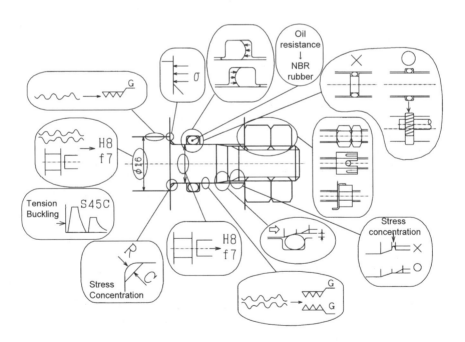

Figure 4.13 Decision process of the mind when designing a hydraulic cylinder

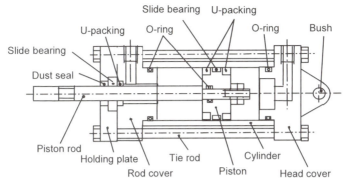

Figure 4.14   Overall structure of the completed hydraulic cylinder

4.2   Designed a Torque Sensor

Project: Designed a torque sensor to detect the force point of action in inserting a polygon mirror with high precision.

Background: An insertion jig helps the polygon mirror insertion (Figure 4.15) and this process takes skill to complete it in a timely manner. Measuring the force during the insertion process will clarify the differences between a skilled technician and an in experienced one. and can eventually aid the insertion process by displaying the information during the process.

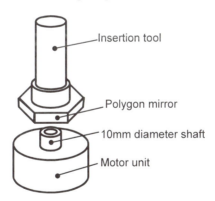

Figure 4.15   Precision inertion operation of the polygon mirror

Motivation: In an earlier model six-axis force sensor proved to have too coarse a torque resolution for finding the force point of action, and we needed to design one with a finer torque resolution. The force component in the axial direction during the insertion process was about 1N, and 0.1N in the horizontal direction. The required torque resolution was 10-5Nm, *i.e.*, to detect a 1mm shift in the point of action of a force of 1gf.

Discussion: To accomplish a resolution of 10-5Nm, we set the rated torque to 1Nm. We then evaluated the basic mechanism and detection method of torque sensing. Plan 1 (Figure 4.16) was to support the structure with a gimbal mechanism and to detect its movement with displacement meters, and plan 2 (Figure 4.17) was to place strain gauges on thin sensing plates.

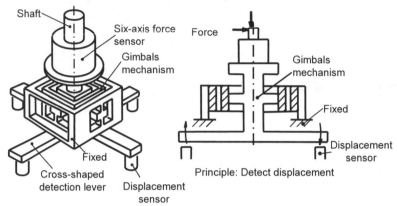

Figure 4.16   Plan 1: Detect torque with displacement

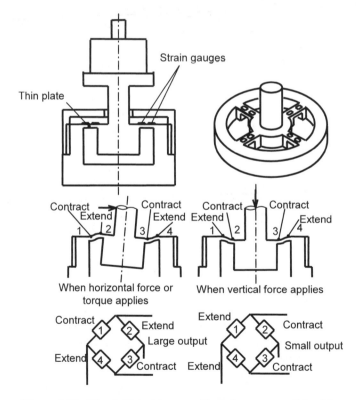

Figure 4.17   Plan 2: Detect torque with strain gauges on thin plates

Alternatives: Using displacement sensors makes the entire structure bulky and we had little experience with them, so we proceeded with the design using strain gauges. For higher sensitivity, we adopted semiconductor strain gauges. As the design progressed, we recognized that plan 2 lacked the rigidity needed to support a load in the axial direction. We then came up with plan 3 (Figure 4.18) to add connection-beams to support the vertical force, support them with cross-shaped beams and place strain gauges on the thin plates of the cross-beams. Eventually we arrived at a design in which strain gauges were attached on the supporting structure we had first planned for the solution with displacement sensors. Figure 4.19 shows the process of the expansion of thoughts.

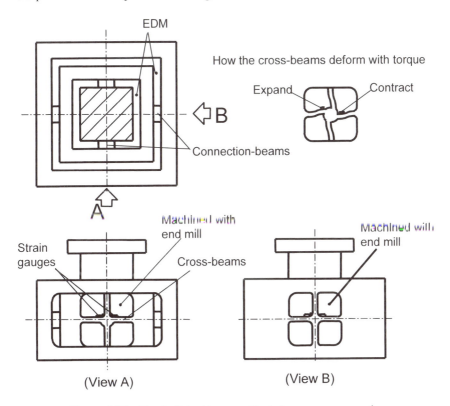

Figure 4.18   Plan 3: Detect torque with strain gauges on cross-beams

Decision: Plan 3 can be machined out, with wire cutting and end mill machining, from a piece of block. We proceeded with this design because of its ease of machining and stiffness. We first planned using 3D CAD, however the structure in the mind was easier to express in 2D, and we drew the design using conventional 2D CAD. If we had had more experience using 3D CAD, we might have constructed the shape with it, however it seems the design's experience and operational skill matter in choosing which tool to use. Figure 4.20 shows the 2D drawing and Figure 4.21 the 3D CAD screen of the structure input based on the 2D drawing.

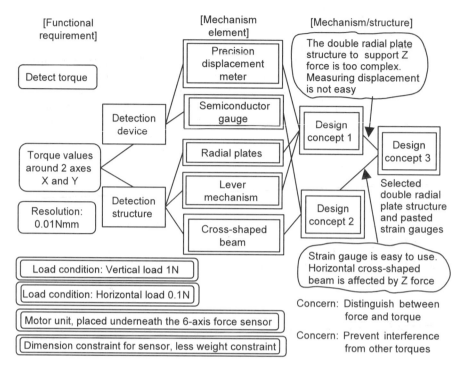

[Functional requirement]

[Mechanism element]

[Mechanism/structure]

Detect torque

Precision displacement meter

Semiconductor gauge

Detection device

Radial plates

Torque values around 2 axes X and Y

Lever mechanism

Detection structure

Resolution: 0.01Nmm

Cross-shaped beam

The double radial plate structure to support Z force is too complex. Measuring displacement is not easy

Design concept 1

Design concept 3

Design concept 2

Selected double radial plate structure and pasted strain gauges

Load condition: Vertical load 1N

Load condition: Horizontal load 0.1N

Motor unit, placed underneath the 6-axis force sensor

Dimension constraint for sensor, less weight constraint

Strain gauge is easy to use. Horizontal cross-shaped beam is affected by Z force

Concern: Distinguish between force and torque

Concern: Prevent interference from other torques

Figure 4.19   Expansion of thoughts diagram of the torque sensor design

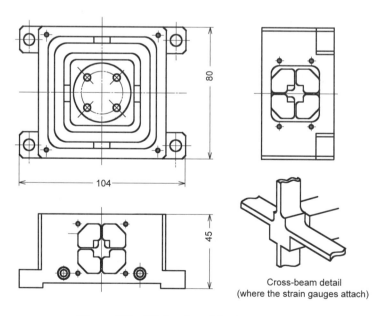

Cross-beam detail
(where the strain gauges attach)

Figure 4.20   2D drawing of the torque sensor

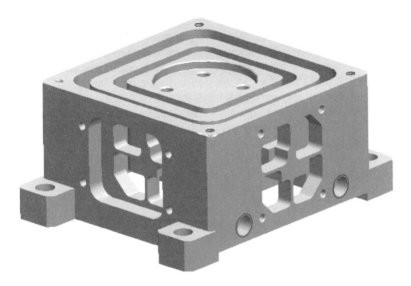

Figure 4.21   CAD shaded rendering of the torque sensor

Production: We sent the 2D drawing to a machining company to have it made. Figure 4.22 shows a photograph of the detection block produced. Comparing the photograph and the 3D CAD image, we were surprised by how closely the 3D CAD image resembles the real object.

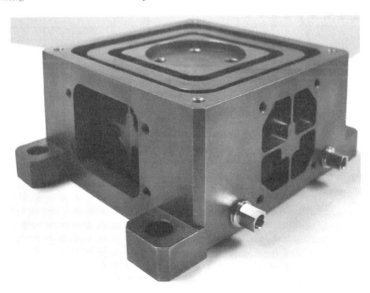

Figure 4.22   Photograph image of the finished torque sensor

Sensor output performance: Figure 4.23 shows the output characteristics of the torque sensor. Figure 4.23(a) shows the response to torque input around the X-axis and Figure 4.23(b) that around the Y-axis. Both diagrams show the main component and the interference. We measured these characteristics by applying a horizontal load at a point with a known height from the sensor center; however, the same measurement with the load applied at a different height produced almost identical results, thus the effect of the load is negligible. The X-axis and Y-axis sensitivities measured, respectively, 0.0043Nm and 0.0074Ms, with interference of 1.2% and 1.4% of the main components. Both curves showed good linearity with hardly any hysteresis. The natural frequencies were 120Hz around the X-axis and 118Hz around the Y-axis.

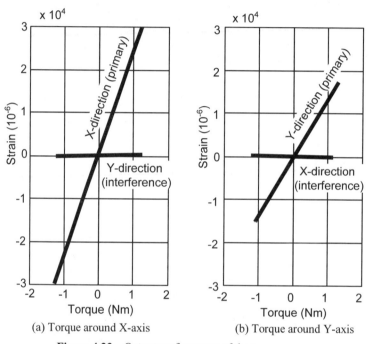

(a) Torque around X-axis   (b) Torque around Y-axis

Figure 4.23   Output performance of the torque sensor

Comparing the output performance and CAE analysis results: Based on the CAD data, we carried out an FEM stress analysis to compare its results with the measured performance. Two pieces of analysis software, DesignSpace and COSMOSWorks, predicted the stress values at locations where the strain gauges were placed. The load conditions were set to the same with the actual performance test with a concentrated load on a bar attached to the sensor face. Figure 4.24 shows an example of the analysis results. DesignSpace showed the torque to produce 100microstrain was 0.0047Nm and 0.0028Nm respectively around the X- and Y- axes, whereas COSMOSWorks gave 0.0089Nm and 0.0082Nm. These values well match the measured values in order of magnitude. Designing with the aid of simulation software and performance testing allow the effective development of machines with higher performance.

Note that getting the analysis results to more closely match the measured results is difficult. Such efforts lack practicality because setting the boundary conditions (constraints and loading) and material property to exactly match the real objects, or refining the mesh for small structures, requires tremendous efforts. The designer should simplify the complex real conditions and look for qualitative reference that resembles about 50% of the real world.

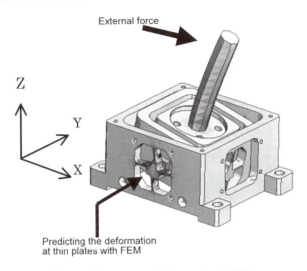

Figure 4.24   Sensor deformation by FEM

Application test: We next took measurements during the insertion of polygon mirrors using our torque sensor. Figure 4.25 shows how our test equipment looked. An off-the-shelf six-axis force sensor (made by Nitta) measured the three force components. As we explained earlier, torque measurement by the six-axis force sensor was too coarse for our purpose. In addition to the force and torque, we measured the orientation of the polygon mirror during the insertion. Figure 4.26 shows the tree-structured expansion diagram for this project. The basic flow goes from left to right. The diagram has numbers to show the sequence of concepts coming to the mind, though some nodes are missing these sequence numbers. Drawing this type of diagram indicates how the mind operated and clarifies the overall picture, and sometimes allows the designer to pick up items left out during the first pass of designing. The diagram serves as a good medium to transfer information to a third party.

Measurement results: Figure 4.27 shows the time history of the three reaction force components when an experienced operator inserted the mirror. From the top are the three components in X-, Y-, and Z-directions. The insertion starts around point a, the Z-component gradually increasing between b and c when the centering takes place; once the center is found, the Z-component decreases. Then section d–e is where the insertion progresses with increasing contact surface and resistance until the insertion is completed at point e. After point e, the X- and Y-components are zero, whereas the Z-component measures the weight of the mirror itself.

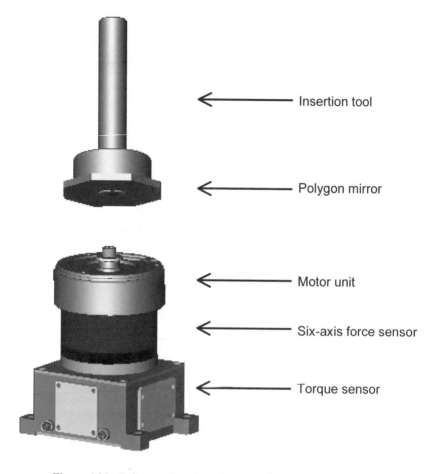

Insertion tool

Polygon mirror

Motor unit

Six-axis force sensor

Torque sensor

Figure 4.25   Polygon mirror insertion operation measurement tool

Both the experienced and the inexperienced operator carried out the process three times, and Figure 4.28 shows both force and torque components from typical processes. Both the force and torque histories show the difference between the two operators, though it is more apparent with the torque. With the orientation information that we also measured, the novice twirls the mirror around the Z-axis to find the center, whereas the experienced operator directly matches the center and without much load in the Z-direction. Measuring the force, torque, and orientation clarifies the process information that is otherwise unavailable from a single piece of information. Revealing the elements and mechanism of how the experienced operator works translates the tacit skills of an individual into scientific knowledge that we can transfer to others and, further, build a supporting system to help the inexperienced.

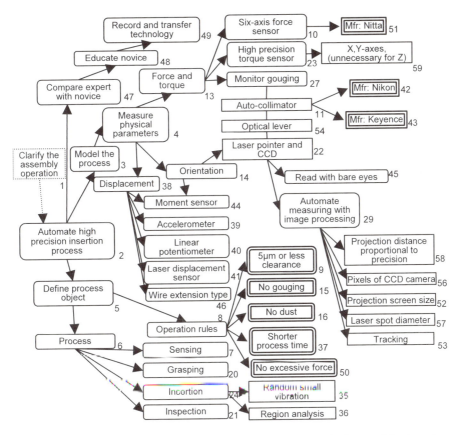

Figure 4.26 Expansion of thoughts diagram for measuring and analyzing precision insertion

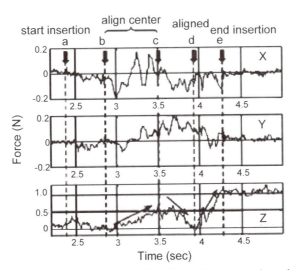

Figure 4.27 Force component time history by an experienced operator

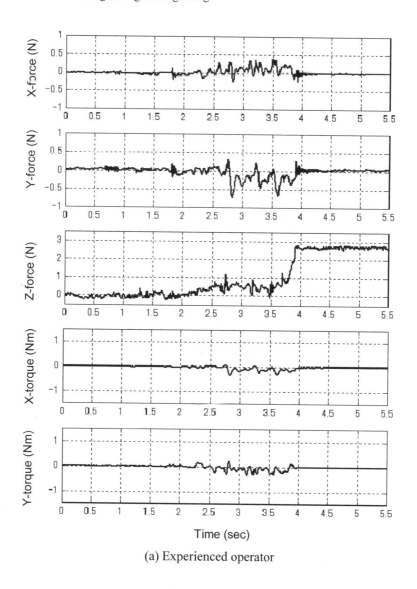

(a) Experienced operator

Figure 4.28   Comparing the force and torque time history between an experienced and a novice operator

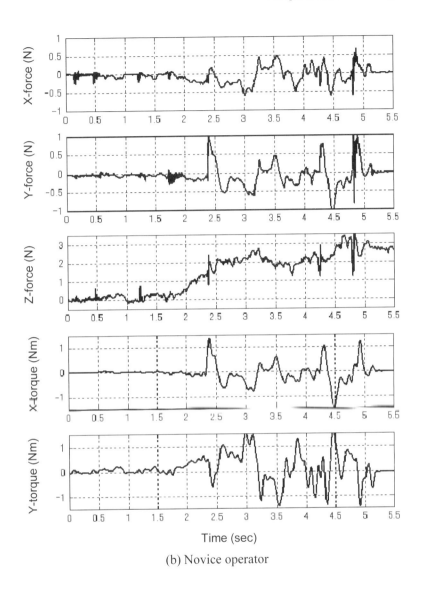

(b) Novice operator

Figure 4.28   *Continued*

My Best Piping Design Turned Out to Be a Disaster
(a note from our group member)

I have been developing automobiles for 30 years and have many experiences that I cannot forget. Let me recount one of them.

It was a year after I joined the company and I was assigned the task of designing the brake-pipe. The conventional design has a number of bends, most of which were 90-degree ones (see sketch below).

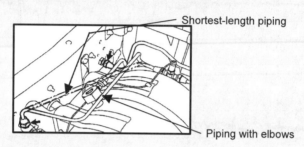

Shortest-length piping

Piping with elbows

"What a poor design!" "These bends cost unnecessary pipe length and bending processes," "The most rational design is to bend the pipe once on the plane that passes the three points that it has to go through," and "Maybe the former designer had no sense of thinking design in 3D." I was so proud of my new design with the shortest length. And the results? Well, it turned out to be a disaster. Many parts in the engine compartment are arranged relative to the vehicle axis and only my pipe took its own direction, standing out in the otherwise clean orchestration. The worst part was my lack of knowledge about a part that interfered with the pipe I designed. It was a result of my designing solely inside my head without knowing the part structure of automobiles. Although I was inexperienced, I experienced the worst failure for a designer.

Through this mistake, I learned that "It is dangerous to act without knowing the object and its surroundings because there is no way to counter what you do not know."

For those who are just starting out in design, the most important lesson is, "Thoroughly study past experience and knowledge."

Author's advice: One of the pitfalls that the designer always has to look out for is avoiding spatial interference. This case is a design error made by overlooking this fact and it is a common mistake for a beginner.

The writer says that he "recognized the pipe running at an angle destroyed the overall balance and was not pretty." This is one of the most important evaluation criteria of design. One of my tacit principles of design is the viewpoint of "function and beauty." It means that something not beautiful has something unnatural or wasteful in it.

4.3 Designed a Positioning Table

One of the basic functional requirements for a machine tool, a measuring device, or a test device is to move a stage in one direction and stop at a specified location (positioning). Design and development of such machine elements is an eternal project, as the required performance keeps getting tighter.

This section analyzes the functional requirements, describes the processes plans to meet the requirements, the mind processes, and decisions in designing the solution.

4.3.1   Design Specification of the Positioning Table

Positioning tables range from large ones like the machining tables of machine tools, or transfer vehicles for products in shops, to small ones like the positioning stages for optical apparatuses such as microscopes. The use largely affects the functional requirement (specification). Specification items include the following (Figure 4.29):
- Length of travel (stroke)
- Size and weight of object to move
- Speed of transfer
- Precision of positioning

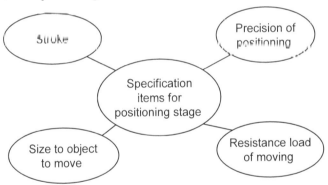

Figure 4.29   Specifications of the positioning table

For the sample in this section, we will set the following values for a small-size machining tool:
- Stroke: 300mm
- Mass of object to move: 10kg; size of load: 100N
- Speed of transfer: 100mm/s
- Precision of positioning: ±0.01mm

4.3.2   Analyzing the Functional Requirements

Given the specification above, here is the list of required functions. First the basic functions are:
(1) Transfer the table in one direction for the stroke.
(2) Guide the movement on a straight line.
(3) Allow no play in the movement.
(4) Provide enough stiffness against loads from other directions.
(5) Move the table at the required speed.
(6) Meet the precision requirement.

4.3.3   Discussing and Determining the Basic Mechanism

Starting by evaluating the functional requirements, Figure 4.30 shows the expansion of thoughts diagram up to the point of evaluating basic mechanisms (mainly the process of selecting the mechanisms) to meet these functions.
(1) Transfer the table in one direction: The driving mechanism can be one that converts a motor rotation into linear motion (ball screw, timing belt, or rack and pinion), a linear motor, hydraulic or air cylinder, and so on. If the stroke is short and the load is small, a supersonic motor, a voice coil motor, or an inchworm mechanism is also feasible. A shorter stroke requirement in the micrometers level can be met by a piezoelectric element, a magnetic strain element or thermal deformation. In contrast, for long strokes, we can choose wheels that travel over rails. From the given stroke, load, and required position, we chose a motor and ball screw.
(2) Linear guidance: Elements for guiding straight motion include linear guides and roller guides. These elements reduce friction by rolling balls or rollers, have no rattling, and withstand lateral load. An element with even less friction is an air bearing. If the load is so big that rolling contact cannot hold, then sliding contact will be the choice. Completely different types that produce linear motion by combining rotating links are the parallel link mechanism, the Scott-
Russel mechanism and others. Here we adopted the most commonly used and rigid linear guide.
(3) Positioning: Methods for positioning include the semiclosed method of counting the motor rotation with an encoder, and closed method that directly reads the table position with a linear scale. The direct measuring method maintains high precision.
(4) Motor selection: AC servomotors have shown much improvement in their performance. This element is now the most commonly used one. The open loop (without sensor feedback) stepping motors are sometimes used as well.

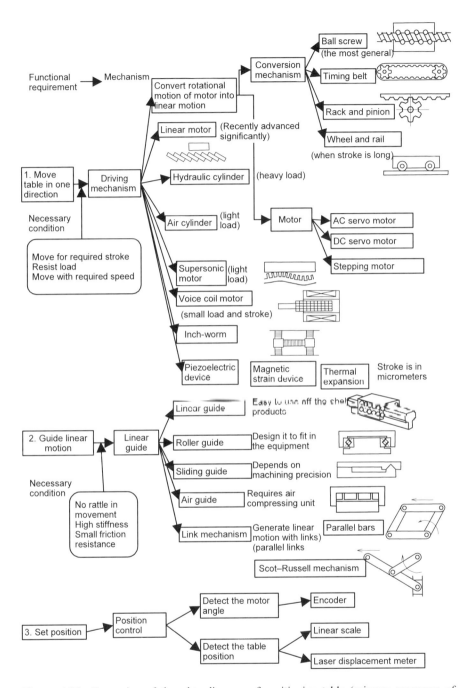

Figure 4.30 Expansion of thoughts diagram of positioning table (primary processes of selecting mechanisms)

4.3.4   Producing the Structure

We have selected the basic structural elements of a motor and ball screw, linear guide, and linear scale; with these we then evaluate the required functions to construct the positioning table and select the necessary elements to produce the overall structure. Figure 4.31 shows the expansion of thoughts diagram for this part. Important aspects to be aware of include the following:

(1) No part must float in mid-air.
(2) The force flow curve has to close within the overall structure when it is still.
(3) The configuration has to be rigid against the driving force or its reaction force.
(4) There is no spatial interference among elements.
(5) The parts are shaped for easy manufacturing.
(6) The arrangement allows easy assembly.

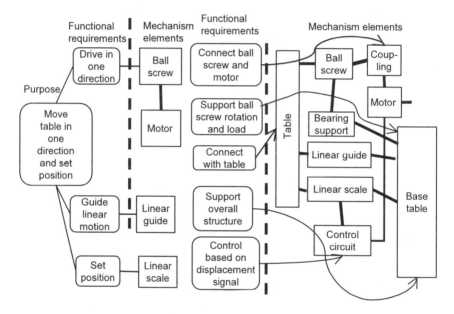

Figure 4.31   Expansion of thoughts diagram of the positioning table with ball screw and linear guide

The ball screw receives axial force as the reaction to the driving force. The driving torque comes from the motor; however, the reaction force acts not on the motor but on one end of the screw. This is because motors are not built to withstand forces in the axial direction. To meet this purpose, ball bearings support the ends of the ball screw. The side that takes the axial force is built specially to withstand this thrust. A coupling connects the ball screw to the motor.

A linear guide placed on the base table guides the table movement and the shaft support that holds the ball screw. We must set the linear scale to detect the movement of the table against the base table. The driving ball screw, the linear guide, and the linear scale all are fixed to the table that is then supported by the base table. Figure 4.32 shows the element components of the positioning table and their relations as they determine the overall structure.

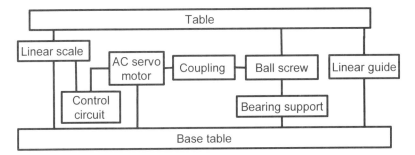

Figure 4.32   Structural elements of the positioning table and their relations

Figure 4.33 is a conceptual sketch of the positioning table. The table has the ball screw underneath it at the center, and linear guides on the two sides. Parallel groves in the base table allow mounting the linear guides in parallel. Pairing the linear guides gives stiffness to the table. For the smooth movement of the table, however, this pair must be set parallel and we installed adjustment screws for this purpose. A linear scale mounted on the side of the table detects the position of the table. Angular ball bearings support the neck of the ball screw (side closer to the motor) to hold the axial force. The other end supports the radial force while it allows movement in the axial direction to release the force from thermal expansion.

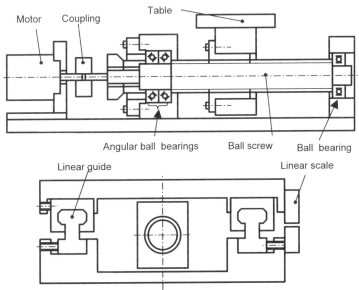

Figure 4.33   Conceptual sketch of the positioning table

4.4   Built an Intelligent Grinding Tool for Producing Flat Wafer Surfaces

Project: Built an intelligent grinding machine that uses sensor information and knowledge for finishing large wafers or LCD glass plates by grinding alone without polishing.

Background: It was the end of 1993. The outburst of general construction scandals changed the flow of strategic funding and, out of the blue, the Ministry of Education was assigned a budget. Our department received an allocation of 4 million dollars. This was due largely to Mr M's effort over 10 years in filing applications for research into machining. After the budget was apportioned, we were assigned half of it for machining tool research.

Route to making a decision: It was my dream to build a grinding machine since the time I was working for a private company. After leaving the company to take a job at school, my dream came true. The dream took 50 million yen paid to T Corporation who helped us build the machine but they said it costed them about twice the amount to finish the project.

It started with only one phone call by Professor H to the president of the company. While I was listening his phone conversation at his side, I learned his scale of "accomplishments, people's trust, dreams, and hoax" was a level different from what I had. We had to announce the project on December 15, collect bids during the next month, and finish it by the end of March. It was a super-express plan (in fact, we finished it on March 90th). Of course, we had to come up with a published specification full of functions that were not available before. The day after we got the oral confirmation, we gathered people from T Corporation, made sure they were going to raise their hands, then, within the following 10 minutes, set the specification while we watched the expressions on their faces. Figure 4.34 shows the sketch we drew then. We had an idea for remote centered compliance (RCC) that lifts the grinding wheel in the other direction when a force acts to tilt it.

Decisions: We first set the functional requirement to "grind an 8-inch workpiece with flatness of 0.1μm." At the time, wrapping and polishing processes produced flatness of 0.5 to 0.2μm and so we set a high goal of 0.1μm. We later found that this 0.1μm was more difficult to measure than to manufacture. We measured the actual product using quartz glass with a flatness of 0.05μm, however the reference itself deformed with the slightest force. Our measurements were made with the thought that "Those who believe are rewarded."

To meet the functional requirement we then set the four functions. Figure 4.35 shows them: "Compensate for the deformation due to grinding force", "Compensate for the deformation caused by heat", "Compensate the feeding error in the transfer mechanism", and "Hold the work without deforming it." Functions to meet the overall requirement did not need to be exactly these four. Other researches, for example, included one that built a structure totally free of thermal deformation by using super-invar material or quartz glass. Yet another had developed a static-pressure oil spindle that did not allow force-induced deformation, or a grinding machine equipped with precision sliding bearings with true straightness. All these other options, however, were expensive. I had seen the grinding machine for the lenses of laser weapons for achieving ultimate precision in the Laurens Livermore Lab in the US, and this visit convinced me that such a project required expensive national efforts just like the time when the atomic bomb was developed.

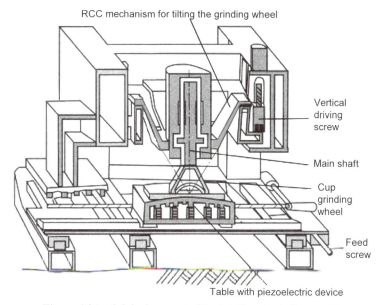

RCC mechanism for tilting the grinding wheel

Vertical driving screw

Main shaft

Cup grinding wheel

Feed screw

Table with piezoelectric device

Figure 4.34   Original concept of the intelligent grinding machine

So I came up with the four functions I mentioned, but they did not come all of a sudden out of the blue. I had made my own calculations. We had already done some prototype experiments to see if we could accomplish such functions. This was our secret, allowing us to suddenly list the functions in 10 minutes and put them in the published specification. My boss Professor H always told me, "Even though you may not have the budget now, always plan the 1 million, 10 million, and 100 million yen researches," and this preparedness saved me this time (of course, 99% of such plans go down the drain).

Since around 1985, Prof. H had been engaged in developing, together with Professor N and Professor M, an intelligent machining center, which compensated the deformation from heat by measuring the column thermal deformation with sensors (made of super-invar material to detect the difference in thermal expansion with the column using strain gauges), and actively bending the column with a thermal actuator (composed of a heater and a water-cooled jacket to produce thermal deformation). For the function of compensating for the feeding error in the driving mechanism, we had already, with the aforementioned T Corporation, developed straight guides and ball screws that enabled fine movement to counter such error using piezoelectric devices. Furthermore, for compensating for the deformation from the cutting force we had already made a number of different six-axis force sensors for measuring cutting forces, and for fixing the work without deforming it we had produced a no-deformation ice chuck. The intelligent grinding machine was an integration of all these technologies we had developed.

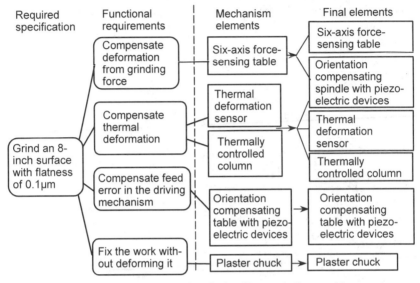

Figure 4.35   Constructing the intelligent grinding machine

If, however, we calculated the precision of compensation of these mechanisms in relation to 8 inches, they were 0.5 to 5μm. Superimposing the four functions we had several micro-meters, so the target 0.1μm was a bluff. Although a higher goal makes us hustle, it was too much, and we later learned our lesson.

Sequel to making the decision: The plan gradually reformed itself as we realized Figure 4.34. The orientation-compensating spindle or table with piezoelectric devices, shown on the right side of Figure 4.34, meets the first three functions. For the deformation-free chuck, after trial and error, we now settled on fixing the work with plaster. Figure 4.36 shows the intelligent grinding machine we eventually produced. It achieved a flatness of 0.3μm/200mm, however we have not yet attained the 0.1μm goal.

Knowledge for setting plans:
    (1) Stay ready to grab a chance, or you cannot do anything when it really arrives.
    (2) Always have relations with people that are willing to cooperate with you, otherwise you cannot make the best use of the chance when it comes.

    •   The goddess of chance has no mercy.
    •   Bluffing backed by a dream has its strength.

(a) Entire machine

(b) Fixture table and cup grinding machine

Figure 4.36   Intelligent grinding machine that we produced

4.5 Planned and Proposed the Nanomanufacturing World Project

Project: We built a Nanomanufacturing World (NMW) that allowed the observation of the object of operation via an electron microscope while working on it using the right-and-left-hand robots. With the help of many people, we finally came up with a system that achieved what we had planned.

Background: It was the spring of 1992. I was astounded when I made my move from a private company to the University of Tokyo. The entire budget (it is called the "school expense") a faculty member receive from the Ministry of education was 2 million yen. My boss Professor H said to me, "Make money yourself. I will show you how." In fact, before I had made a move, he whispered in my ear, "You can do whatever you want to in the university." For a corporate engineer that is controlled from what job to do to where he works, these were the words of a dream. Nonetheless, not stating the condition that "only if you have the money" was the trick. I, however, had thought that research funds would always follow the most famous University of Tokyo. I could conduct research with the money that my boss made, but the money was accompanied with orders. Just the spirit alone by no means accomplishes independence.

Route to making a decision: First I had to put up a sign. A sign that signified my "specialty". I was troubled by the question, "What is your specialty?" that everyone asked as soon as I made my move. While I was in the United States, I only worked as a field engineer and that alone was nothing special.

Right around that time, "micromachines" were catching the attention. When an American researcher applied the manufacturing methods for semiconductors to producing silicon gears and turning electrostatic motors, it turned into a boom in the US. The Japanese Ministry of International Trades and Industry (MITI) judged that this was the seed of a future industry and started to promote its research. I tagged on to the trend. If you wanted to fish, you'd better pick a pond where there are some. I was proud that I had done my fish hunting first, but when I look back at things, I realized that the government had already set the framework for letting the funds flow. In fact, after this, I never received a single drop of funding from MITI, but at the time I never saw what was going on backstage.

Our proposal was not one where we ventured into a different field. Professor H had already produced a manipulator called a "nanorobot" that moved the tool using piezoelectric devices under the vision of an electron microscope, and we knew what research we had to do. So-called virtual reality, for example, was needed for the human operator to work with microtools as though he had them right under his nose. We had to fill out our application form with results from preliminary experiments, and these later expanded into my dreams, as the following paragraphs describe.

Thoughts for making the decision: Next, I wrote stacks of funding applications. The forms had to reveal our sincere pledge for money between the lines. Filling out such forms, however, was not easy. Boasting that I had spent millions of dollars on research during my nine years of working as an engineer was of no use this time. In the university, I needed the experience of writing applications for charity money that we had no intention of returning, which was quite unlike asking for a loan or investment.

When you are seeking charity money, the most important thing is to talk about your dream (Figure 4.37). It is much like an artist sweet talking a patron. Singing the wonders of what happens if the study goes well, and whispering that you are the only key person that can help this one and only bet. The one that betting the story also finds his mind rising with the phrases and starts floating with thoughts that dreams have already come true....

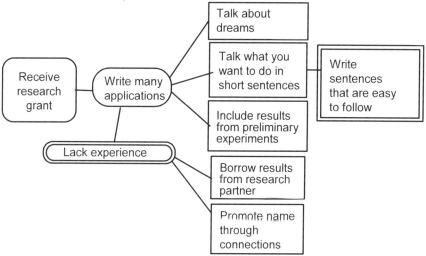

Figure 4.37   Filling out grant application forms

Corporate management in Japan dislikes engineers that speculate. No matter how extraordinary the technology you explain is, they always ask "How many years would it take for it to make a profit?" Thus, the longer you are in the corporate world, the fewer dreams there are in your writings. Every time I finished an application, I had Professor H pretend that he was judging the file. "This is no fun!" he repeated, and I had to edit it over and over. On the other hand, sentences too well written lack real enthusiasm and I learned to leave them just at the right level.

Another important factor is, "Keep it short to explain what you want to do." A businessman, if he bumps into the president in an elevator, has to explain his work before the car reaches the president's floor. A funding application is the same. In general, a civil grant requires several pages, and a government grant several tens of pages. Nevertheless, you still have to keep the words short and with modulation. This is because tedious sentences that talk about possibilities or adequacy just bore the great authorities that have to read them. If the first three minutes of an application catches his attention, then the great authority will spend another three minutes reading it. Even so, the great authority will have to spend a whole day to go through 100 files of applications, and furthermore, only a few of the great authorities are asked to judge for all the different grants, and those end up with hardly any time. No matter how thick your application is, you need to make a quick impression that this is something good from the very beginning of the file. If then the contents are hard to understand, the reader will not get excited, so you need to compose sentences that instantly work their ways into mind. Writing about exceptions or what assumptions are made, like you would for a thesis, has no effect. Pages with small characters and without concise paragraphs lose the reader's attention. The truth is, I secretly had one of the great authorities teach me these tips (and they are of benefit for working at the University of Tokyo, which has a number of these great authorities in their faculty).

The phrase "Nanomanufacturing world" is one coined by Professor H. Although the phrase is short, it sounds full of dreams that make you imagine about a tiny world. He said that he settled on the word "world" from "It's a small world" in Disneyland, after changing "micro" into "nano", and trying out "space" instead of "world". It was much like composing the lyrics for a pop song.

Nevertheless, what mattered the most then were the results we had so far. These are like credibility in the business world. Venture businesses don't just lack money, but with no credibility they cannot even borrow it. If the judge glances through the authors and thinks "I don't know these guys. Next!" then that's the end. Some quote the words "chicken or egg;" however, for a newcomer there is neither chicken nor egg and that makes the problem worse.

Under such circumstances, the only way out is to rely on personal relations to win credibility, or use connections. In other words, have someone with established results join the team or sell your face to the judge by other means. Young researchers in America are not afraid of challenging the great authorities at social gatherings to have them remember their names.

Sequel to the decision: In the end, we won a three years' grant of 32 million yen from a civil foundation the TEPCO Memorial Research Aid Foundation (This one was limited to those under the age of 40 and I was able to compete without experience.) And then came a three-year scientific research grant of 26 million yen to start from the following year from the Ministry of Education, and a 20 million yen annual grant from E Corporation (we borrowed Professor H's name to head the research efforts). We even had E Corporation build the Nanomanufacturing World shown in Figure 4.38 (we also had to borrow the credibility of Professor H. for this as well). We later learned that E Corporation spent about 200 million yen to do this. Figure 4.50 in Section 4.7 shows the expansion of thoughts diagram of the fine-positioning mechanism for this system. We titled the papers we wrote about our work "Construction of Nano Manufacturing World" and they were accepted in Europe and the US. We then wrote a paper with the same title (in Japanese) for the Japan Society of Mechanical Engineers who turned it down saying "Nano is an academic expression meaning ten to the minus ninth power." We later thought this over and wrote a paper in thesis-like sentences which was then accepted by the Society of Electrical Engineers.

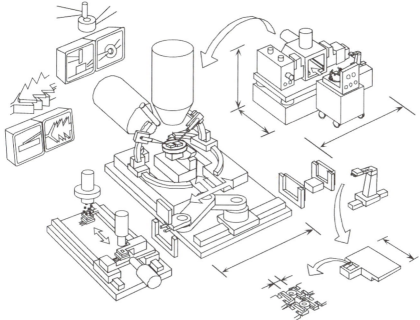

Figure 4.38   Structure of the Nanomanufacturing World (joint research by the University of Tokyo and Ebara Corporation)

That was in 1995. Since then we have produced some results. Building on what we have accomplished so far, we try out some ideas and write "This is what we want to do" in the applications. Once we receive the grant, we write a report "This is what we have done" in the past tense. During this time, we try some new ideas and store the results for later use. In other words, our results are ahead of the funds and our research activities expand with the multiplying effect.

Even now, we write applications on top of stacks of rejected ones. Unless we shoot the bullet, we will not hit the target. Starting in 1996, we have applied for scientific research grants totaling 100 million yen and got 20 million yen each year. In 1997, we tried for a grant from the Ministry of Health, though we were really outside of its scope, and won 80 million a year for 5 years. Composing a grant application in the middle of the night just before the deadline produces some very wretched feelings.

Knowledge for making decisions:
    (1) The applicant has to talk about his dreams to the patron.
    (2) You have to write what you want to do in short paragraphs.
    (3) Newcomers need to borrow results and use connections.
    (4) Seeds do not grow unless they are planted. Do not let rejections discourage you; keep writing the forms.
    (5) Always keep a sharp mind in order to write an attractive proposal.

4.6   Developed the Control System for Automatic Grinding of Turbine Blades

Project: Developed an automatic grinding control system (Figure 4.39) that allows arbitrarily changing the machine tool compliance (the inverse of rigidity) for automatically grinding a turbine blade edge to the desired shape with high precision.

Figure 4.39   Turbine blade edge grinding machine

Background: The process of grinding turbine blades, especially forming their edges into desired shapes, required the human operation of pushing the work against grinding wheels while sensing a subtle change in the grinding force; this process was not yet automated and lacked high precision. A project plan arose to automate this process. The author, on the other hand, had research experience in force sensing and compliance control for improving the precision of machining. The idea of applying compliance control was a good one and we settled to develop a control system for automatic grinding.

Functional requirement: We watched the actual process of grinding the turbine blade edges in a shop. The operator, as Figure 4.40 shows, held the two ends of the turbine blade with both hands and pushed it against a rotating grinding wheel. He moved the blade sideways and as he turned the blade he gave the edges their desired shapes. The pressing force measured about 700gf (6.9N) and the arms held the work gently as they remained bent.

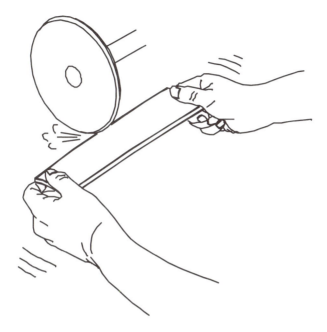

Figure 4.40   Manually grinding a turbine blade edge

We collected the following functional requirements through watching the process. If we can realize this machine, it should be able to hold the piece gently just like a human operator did when he held the blade.

(1) Without pressing the blade firmly, the mechanism should hold the blade softly as with a soft spring, and the spring compliance should be freely changeable.
(2) The grinding would proceed in the long direction of the blade and in the plane perpendicular to this direction, grinding points where the grinding wheel hits the blade should be specifiable.
(3) The force pressing the blade onto the grinding wheel should be arbitrarily set.

The natural spring compliance worked in the positive direction that deformed the system in the direction in which the force was pressing. We, however, configured this machine to produce negative compliance as well, which deformed the system in the direction opposite to the direction of the force. This allowed compensating for the deflection of the grinding wheel from the grinding force and realized high-precision machining with no machining error (see the appendix at the end of this section for details of high-precision machining using negative compliance).

Deciding the mechanisms and structure: Figure 4.41 shows the process of selecting which mechanism to use to meet the functional requirement, and Figure 4.42 the composition of the mechanism (structure) that we selected.

The first big decision to make in the process in Figure 4.41 was whether to adopt a passive method with elements like springs, or to go with the active control method. The passive method would have required springs with variable stiffness and mechanisms like variable dashpots that seemed to likely produce greater complexity, we therefore set our minds on the active method.

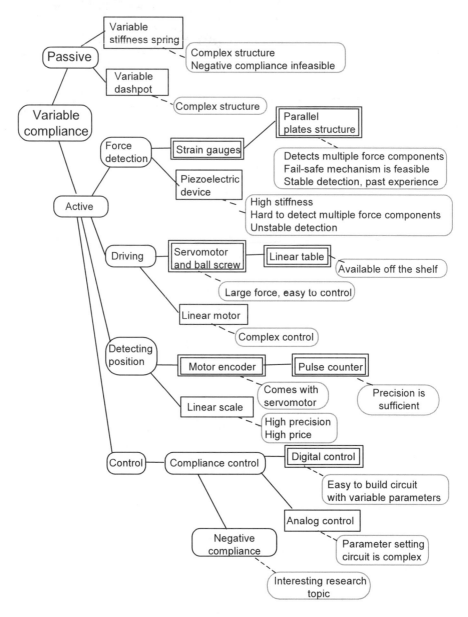

Figure 4.41   Process of selecting the mechanism

To produce variable compliance with active methods, we needed to measure the machining force on the blade and control the blade position accordingly. Methods we needed to decide were those of force detection, driving, position detection, and control. We developed a force sensor with a parallel plates mechanism and strain gauges for detecting the force. For the driving mechanism, we bought an off-the-shelf straight table with a servomotor and ball screw (something is available, buying it proves less costly, after all). Position is usually detected via a linear scale attached to the movable table, however the servomotor to drive the table had an encoder that generated pulses as the table turned and counting those pulses provided sufficient precision. A piece of PC software conducted digital control to produce variable compliance. The software made it easier to change parameters, simplifying the electrical circuits needed.

Figure 4.42 shows the structure of the turbine blade grinding machine built with the decisions above. The turbine blade was attached to the force sensor via a holder, and the force sensor was mounted on an X-Y table that moved in the XY plane. The grinding wheel was turned by the spindle motor and the Z-table moved it vertically.

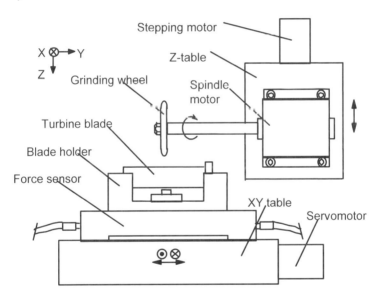

Figure 4.42   Construction of the turbine blade edge grinding machine

Control system structure: Figure 4.43 shows the information flow to accomplish variable compliance. The system drives the table in the target direction calculated from the machining force sensed by the force sensor and the current table position from the position sensors. For example, through setting the origin to the table positions when there is no force, and moving the table by 1cm when the grinding force is 1N and 2cm when it is 2N, the system produces a spring constant of 0.01m/N (=spring stiffness of 100N/m).

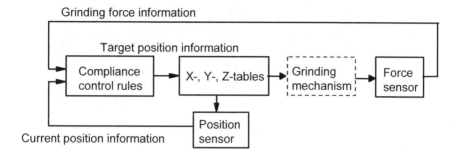

Figure 4.43   Flow of variable compliance information

Figure 4.44 shows how we constructed the compliance control system. Multiplying the grinding force to the compliance characteristics produces the target table position. The system then compares the target with the actual table position and amplifies the difference to drive the table. The loop generates a speed command by magnifying the position error for v/f conversion (the v/f converter transforms voltage to frequency), and drives the table by sending pulses through the pulse driver to the servomotor. The encoder on the servomotor gives the motor angle of rotation, thus counting the pulses represents the table position. External noise that would cause deviation in the actual table position from that given by counting pulses includes the play in the ball screw; however, such error is negligible.

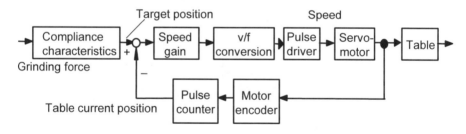

Figure 4.44   Construction of the compliance control system

Dynamic performance: We evaluated the system's compliance control performance to see how our variable compliance control was working. Figure 4.45 shows the table position when force was applied in the X-direction with different compliance values set in the X-direction.

Compliance control performed the following:
(1) Positive compliance → Moved away in the direction of the force like a regular spring.
(2) Zero compliance → No movement even when force was applied. High stiffness.
(3) Negative compliance → Moved in the direction opposite to the force, a negative spring.

Figure 4.45 shows the changing the controller setting produces different types of compliance including negative compliance that moves the tool in the direction opposing the force direction.

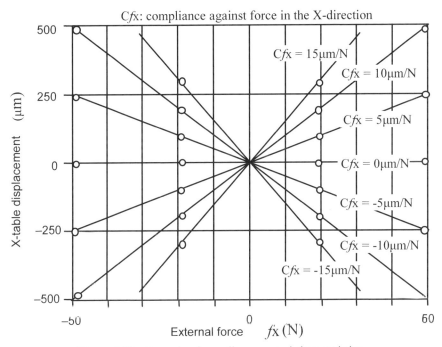

Figure 4.45   Example of compliance control characteristics

Figure 4.46 shows the results of our grinding experiment. We intentionally placed a 50μm bulge and ground it with different compliance settings. Figure 4.46 is the surface form after the grinding. The compliance setting is normalized to the grinding mechanism setting.

When the normalized compliance was 0, the process left about a 25μm bulge. This was due to the softness of the grinding wheel, the deflection of the work caused by the grinding force and other properties of the system. Setting the normalized compliance to a positive value left an even bigger bulge. This situation was equivalent to holding the blade with a soft spring that moved away when the grinding force was applied. A negative normalized compliance left a smaller bulge. In this case, the blade moved in the direction to counter the deformation of the structural components in the direction of the cutting force. Note that setting the normalized compliance to −1 left almost no bulge. This condition exactly countered the deformation of the grinding wheel and other structural components.

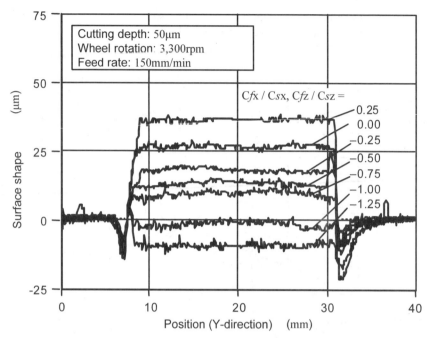

Figure 4.46   Effect of compliance setting on surface shape

Summary: The goal of developing a grinding system that allowed arbitrary setting of the compliance was met. An active scheme had the upper hand over the passive scheme because the former was easier with greater degrees of freedom.

In fact, when building a large-scale system, which scheme to apply and to what extent is an important judgment to make that affects both performance and cost. Control circuits used to consist of analog and logic devices, but recently the trend has been to put the sensor signals through A/D conversion, process the signals in a CPU, and drive the machines with D/A converted signals. This digital scheme costs less and is more flexible in terms of design changes.

Sequel: This automatic grinding system was a technical success; however, it did not make it to the actual shop floor. One of the reasons was that there were further items to develop, like automated chucking, dust proofing, a fail-safe mechanism and so on. Above all, however, the idea of applying automatic control to a subtle machining process faced hurdles in the application of the technology to the field. Sometimes, putting technology elements into practice takes more effort than developing them.

Appendix (High-precision machining with negative compliance): Figure 4.47 exaggerates the machine deformation during a regular machining process. Machining force, as Figure 4.47(a) shows, is generated at the point of contact between the tool and work. This force flows, as shown by the C-shaped curve, from the tool, through the machine to end at the work. Figure 4.47(b) shows the deformation caused by this force. This deformation follows the machining force C-loop and causes machining error at the machining point. The conventional method of countering this error has been to raise the stiffness of the machining system by using thicker structural components of larger bearings. This solution produced the problem of larger, heavier and more complex machines to accomplish high-precision machining.

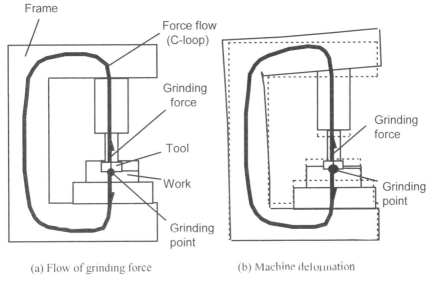

(a) Flow of grinding force   (b) Machine deformation

Figure 4.47   Machine deformation caused by grinding force

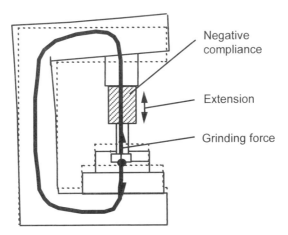

Figure 4.48   Compensating the deformation with negative compliance

Figure 4.48 shows the method of applying negative compliance to compensate for the error. By inserting a device to generate negative compliance somewhere in the C-loop, the device actively expands or contracts to negate the deformation of the machine tool. Selecting the right compliance size then can accomplish zero machining error. This method enables high-precision machining without stiffening the machine tool.

This article is a good example of the mind operation, inverse, and exchange. Negative compliance in this example is derived from inverse operation. It is a result of thinking if some idea can produce deformation in the direction opposing the force, which usually produces deformation in the direction it acts. The idea was that it may produce a desirable effect that otherwise does not happen. This example proves the effect of thoroughly thinking about a suggestion from inverse operation of the mind.

4.7   Built "Creative Design Engine" for Assisting Idea Generation

Project: The designer will be greatly helped by a system that displays the knowledge needed for the design, and the process of mind decisions. These processes are important in knowledge management and we actually built a system which we named "Creative Design Engine."

Background: The authors had published *The Practice of Machine Design, Book 3 – Learning from Failure* (Yotaro Hatamura eds., The Practice of Machine Design Research Group, Nikkan Kogyo, 1996 and *Knowledge Management of Design* (Masayuki Nakao, Yotaro Hatamura, Kazutaka Hattori, Nikkan Kogyo, 1997). In the following year, 1998, we came up with the idea of developing the software "Creative Design Engine" that displays knowledge about those failures and idea generation during design. The idea gained great attention, though unfortunately only within our group. When we thought it was going to end as just an idea, the Japan Mechanical System Agency supported it and decided to fund our research.

Route to making a decision: In addition to the contents of the above books, we gathered together 11 corporations as participants and collected and edited almost 1,000 cases of design errors and ideas generation for design. We also configured the application so that it could search the verbs that expressed the design project in addition to a noun-based search. We added a thesaurus to look up related words to keep the string of thoughts going during the design session. The tool, however, did not work as we expected when we had the students try it. They could not find the knowledge needed, while some could not even type in the proper words in Japanese. Something was wrong.

Thoughts on decision-making: Not knowing what you want to know is equivalent to not knowing what you want to design. If that is the case, then,we thought it would help to visualize the mind process of design.

The mind process of design is to arrange the design project or functional requirement (and constraints), solution, functional elements, mechanisms, and structure in words. Putting them in a diagram in circles or boxes and connecting them with lines would visualize the process. Figure 4.49 shows an example.

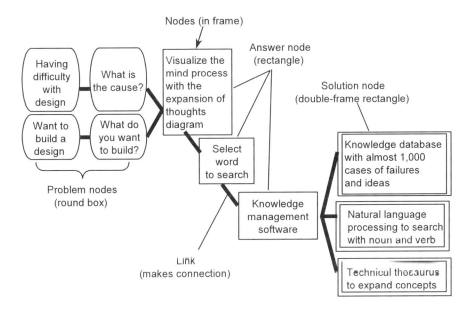

Figure 4.49   Sample session from creative design engine

We had assumed the design solution that "If you can specify what you want to know, then you can search the knowledge." This is the same as the witty monk Ikkyu saying that "I will catch the tiger for you if you can chase it out from the picture." The tiger never comes out. The fact is we needed to perform the step of chasing out the tiger to "Clarify what the real cause is and what you want to do by realizing the design you are wondering about." So, to fill the missing piece in first conceptual level, we added a piece of software to draw the expansion of thoughts diagram in the Creative Design Engine. Figure 4.50 shows what kind of expansion of thoughts diagram one can draw on the screen. This diagram proceeds from left to right with the project on the left, followed by the functions required, then their solution to support visualizing the mind process. When the designer advances his ideas as he constructs the expansion of thoughts diagram on the screen, the diagram clarifies, in an orderly way, his problem of what he has to accomplish and what he can use to solve it. The example in the figure is the expansion of thoughts diagram for the fine-positioning mechanism we used for the Nanomanufacturing World described in Section 4.5.

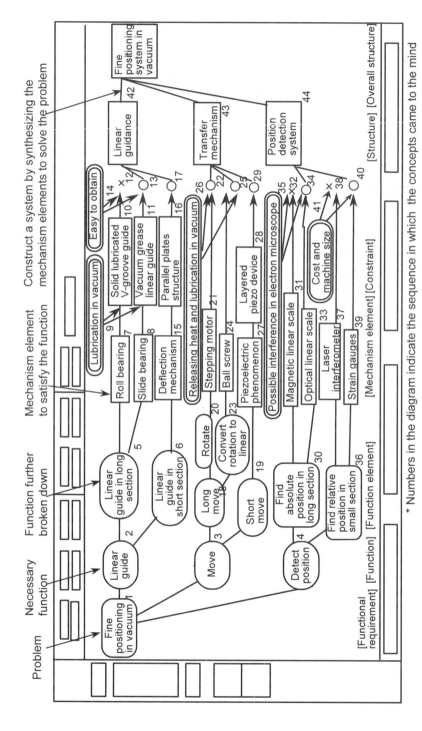

Figure 4.50 Expansion of thoughts diagram displayed by Creative Design Engine (for the fine positioning mechanism for Nanomanufacturing World)

The designer first describes the design project in his own words (what problem he has or what he wants to design). For a young engineer, the phrase is often not very technical, and by clicking on the thesaurus on the screen, engineering terms that relate to his phrase appear. The designer then picks the words that he thinks match his purpose and places them on the expansion of thoughts diagram screen.

In the next step, of extracting the functional requirements to meet the project goal, the designer again picks the related words with the help of the thesaurus and customizes them to appropriate expressions for his project. The user needs to keep in mind that the computer is like a dictionary and the person who analyzes the function requirements to list them up is, after all, the designer himself.

The thesaurus also helps the search for the mechanism that realizes each function listed. That is because we constructed the thesaurus database so that it relates mechanism elements and solutions to each functional requirement. When the screen shows a design example, clicking a button on screen opens an actual design that realized the required function. The designer can read about the design example to learn about issues and mechanical elements that relate to his design. Creative Design Engine supports the designer in the tasks of organizing what problems he needs to think about, expanding thought from function to mechanism, and gaining the necessary design knowledge.

To add to Creative Design Engine we also developed the "behind-the-scene" software function. This is where each former designer can record what alternatives he had, what he thought about when he made the decision, and what results he saw with the design. Such design information is usually not written anywhere, but it is what the follower designer wants the most.

Sequel after making the decision: We had some schools and corporations try out this software for drawing the expansion of thoughts diagram. At schools we had the students draw the diagram on paper (for lack of sufficient PCs) and we found that regardless of which school they graduated from or what design experience they had, only half of them successfully drew the diagram without much guidance. The remaining half, moreover, could not draw it even after we tutored in them in how to do it.

At corporations, about 80% of the managers drew good expansion of thoughts diagrams. If anyone could not, he would not have been promoted to the position. If the manager could not explain to the group what he wanted done, it would be difficult for the organization to get anywhere.

Maybe our ability for logical thinking is formed when we are in grade school. I would, however, like to find a method of tuition to enable anyone taking the course to draw the expansion of thoughts diagram.

Knowledge when deciding about a system: Without the ability to analyze the vague requirement of what the problem is or what one wants to produce, we cannot expect to make use of past knowledge acquired by others to find our design solution. Without training such ability we cannot make use of knowledge management software.

Boasting about our own design solution narrows our vision and solidifies our minds. We will then start to lose what the original design project was about. The expansion of thoughts diagram is effective in visualizing our own mind process. It helps us recognize how our thoughts are structured, what is missing, and what we should think about now. It also helps in transferring our own idea to others.

4.8 Guided Students' Free Imagination in Building Stirling Engines

Project: A regular engine is an internal-combustion type that burns gasoline inside the cylinder and uses the heat to generate force. A not-so-well-known engine type is an external combustion type that heats gas outside the cylinder then brings it into the cylinder to generate force.

We gave the students the assignment of designing and prototyping Stirling engines, and were surprised by the large variety that they came up with.

Background: It was in 1979, when I was a junior, at a lecture about machine design. The lecturer was Professor N who, at the beginning of the class, opened his bag to take out an odd-shaped object. It was a hand-made Stirling engine, with a structure like that shown in Figure 4.51, which, to my amusement, started making a rattling noise as he warmed the bottom of the chopstick can with a methanol lamp. He explained what caused the engine to turn with a PV diagram, but it was beyond my comprehension. In particular I had no idea what the displacer, with a 90-degree phase shift to the piston, had to do. I hated thermal engineering, which uses all the partial differential equations in the world, and my brain simply refused to even think about PC diagrams or ST diagrams. The motion, however, caught my attention. How amazing! Once the lecture was over, I asked the teacher if I could build the same thing.

He advised me to use a honed pipe, and I simply followed it and went all the way to a small machine shop in the town about an hour away. 17 years later, I realized it was an important piece of advice, but at the time I did not even know the grinding process of honing and I simply did what I was told. I borrowed a lathe in the school shop to machine a piston to fit inside the purchased pipe, made holes with the drill press, fastened the parts together with screws, filled the piping gaps with silicone for bathtubs and three days later I had my own finished engine. I then filled the reservoir with water, lighted the methanol lamp, and after a minute, gave a push to the flywheel and then it started to turn making the same rattling noise. What a sensation it was! It kept turning for about 20 minutes until the water in the reservoir turned warm. I put a rubber band on a motor shaft and pushed it against the flywheel to generate power. It could not turn a light bulb on but it successfully lit up an LED.

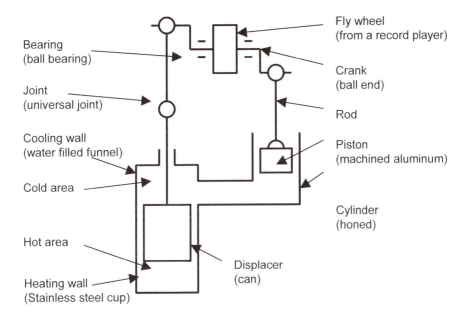

Figure 4.51   Structure of a Stirling engine

I was very proud and brought it to the school and Professor H, who was teaching a manufacturing class, asked what that thing was. So I took it to his lab and with a smiling face, showed how I could turn the engine. Professor H then said, "Ah, the displacer is taking the role of the little boy switching the valves." What little boy? He then started to draw a sketch and explain the mechanism, "First the heat expands the air volume to lift the piston. To continue the motion, it needs to lower the piston by switching the valve and introducing cold air. But there is no valve, so instead the engine lowers the displacer to bring air into the low-temperature part and send the cold air to the cylinder. When the piston has made it to the bottom of the stroke, it then lifts the displacer to bring air into the hot area to warm up the entire volume of air. The warm air then enters the cylinder to lift the piston," he then continued, "That is, the boy who had the task of switching the valves wanted to take a break and invented the displacer to attach to the crank and perform the role of switching the valves."

It was then that I realized, "Well, that's what was going on."

Professor H then handed me a book to read. It had a story about a boy who made a steam engine run by itself. That was his secret source book! I then read on to learn about Newcomen's engine, Watt's invention of the condenser, and Carnot's derivation of thermodynamics, which took these geniuses 100 years of trial and error. Although the engines in the 1750s had efficiencies of about 1%, they brought the Industrial Revolution. Historically, theory followed practice, and success followed failure. I thought it is questionable to teach students, who had never touched machines, the beauty of the theory with entropy, tensors, Navier–Stokes and so on.

Path to making a decision: Years went by, and 17 years later in 1996, after making some career moves, I, who used to be a student, took a teaching job at my school and started to plan an exercise class for junior students. The exercise could have been anything but there had to be motion for the fun of it. Then why not have them make a Stirling engine? It only takes three days. I did not really think hard and made a big decision, but rather it was without much serious thought, and I thought, "It will turn out whichever way it will." Maybe that is why the project lasted for years.

Thoughts when making the decision: Telling the students to make a Stirling engine will lead them nowhere. When I had made one 17 years before, I had just followed what Professor N had. Even so, it touched my heart, so why not just have the students follow what was available? Professor K told me he had a drawing used at a different school in an exercise. Well, why not use it? I started the project with a completely Japanese method. I sent 20 students to the school's machining shop in groups of 4 to prototype a model within 10 hours. I motivated the course assistants by promising a 20,000 yen bonus if all 5 groups succeeded, 10,000 if 4. The success rate turned out to exceed 60%.

Besides this seminar, I organized another seminar that brought in all kinds of engines models from an off-the-shelf internal-combustion engine to a steam engine, studied them and mimicked them to build our own. After a year, the exercise gave us insight into what was important in designing small Stirling engines. The primary function of thermal engines were written about in plenty of books so that was not too difficult to accomplish. We found more problems in the secondary functions; which I will describe later.

Until 2001, the exercise class kept evolving by showing further creativity. The students stopped mimicking existing models, and started to take on creative designs that they would boast about (Figure 4.52). This advancement has been due largely to the contribution by Professor A who specializes in engines. He never turns down off-the-wall proposals by the students and provides good guidance for them. The students then started to design their own models. The level the students have reached has now surpassed the teachers (at least myself) and their presentation just leaves me speechless.

The students, however, did make mistakes. Their failures had patterns. Without experience in manufacture, what they thought about in their minds sometimes did not come true. This is where the teacher comes into the scene. Figure 4.53 shows the expansion of thoughts diagram for instructors when they go around patting the shoulders of students who are in tears because their machines will not turn. All tips in the diagram are not in the thermal engine textbook and are not related to the primary functions. A professional designer would take a glimpse and say, "Oh, that is common sense," but these are mistakes about the basics of design that any novice designer would make; and I made plenty of such mistakes earlier in my career. This is the King's path to design, scenario of failure, and shortcut to success. I actually printed out this diagram to give out during the first class, but the students still made the same error. Probably the knowledge-seeking part of their minds does not activate unless they make the mistakes themselves. Even though it is the King's path, the brain only activates when shaken by necessity to seek appropriate knowledge.

(a)

(b)

(c)

(d)

Figure 4.52   Stirling engines produced by students

Let's go back to the diagram. Failures are categorized into "Poor manufacture" and "Poor design." The most common poor machining is in the process of cutting the cylinder hole. Straightness, circularity, and surface smoothness are all hard to accomplish and the result often freely lets air past the piston. It is very difficult for an inexperienced operator to produce a piston and bore mating with a gap within 0.01mm and a surface roughness of 0.001mm on the bore of a 20mm diameter cylinder. Furthermore, the students had to work with the outdated manual lathes. Japan was late in developing its own boring technology and during the World War II the engines were always built by mating on the spot, thus parts could not be replaced in the field. It is not hard to imagine the difficulty the students had to go through. So, it is better to use honed pipes, syringes, or dashpots. Next is poor sealing. We can detect this by submerging the unit underwater, turning it and seeing if air bubbles come out. If so, high-temperature glue is put over the holes. If the piston was made too loose, we can spray WD-40 inside the bore to produce a thin layer of oil. These failures teach the students about the importance of rubber products like O-rings or U-packing.

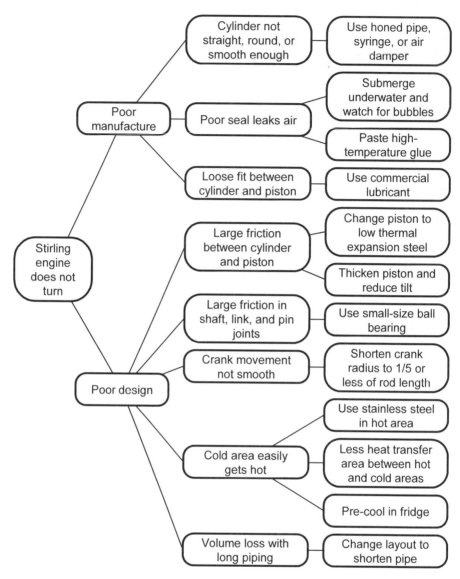

Figure 4.53   Expansion of thoughts diagram for the instructor in Stirling engines

The solutions so far have solved over half the problems. We are left with poor designs. First comes the gouging of the piston caused by thermal expansion, followed by plastic deformation and increased friction. If the honed pipe is made of steel and the piston of aluminum, the difference in the thermal expansion coefficient is "1cm, 1degree Celsius, 1μm." If both are subject to a temperature rise of 10degrees Celsius, an aluminum disk with 1cm radius will expand inside the steel and the 0.01mm gap will be gone. Naturally the piston gouges into the cylinder bore. This problem occurs more frequently with double-piston types, instead of the displacer types shown in Figure 4.51, which have the heated area closer to the cylinder. Making the piston lighter has its advantages, but the designer must keep an eye on how it performs in its mating cylinder. The next problem is friction. Using slide bearings for rotation and other joints produces friction that can no longer be ignored. In this case, off-the-shelf small ball bearings, like those for 4WD model cars, work well. Another mistake is a crank radius that is too long. The torque cycle gets too large to interfere with smooth rotation. Professor A recommends keeping it shorter than 1/5 of the rod length.

In designing small thermal engines, the biggest problem is the heat transfer that quickly evens out the temperature difference between the hot and cold areas. If you solve the equation of heat transfer, you can soon see that the time for heat transfer is proportional to the distance squared. In other words, if you make the engine 10 times smaller, the heat travels in 1/100 of the time. A thermal engine stops working when the temperature distribution evens out. Solutions are to keep the heat transfer cross-sectional area between the hot and cold parts small to minimize the heat flow or to use a material with a small heat transfer coefficient like stainless steel. If your deadline is approaching and you have no time for a remake, you may want to substitute the water in the reservoir with ice or to cool the entire engine in the freezer before your test. Keeping the piping short to reduce the dead volume is also effective. Piping, just like wiring, is often ignored because it does not seem to affect the primary function, however, if pipe routes are too long, the dead volume is no longer negligible.

Sequel after making the decision: We built a micro-gas turbine with a rotor diameter of 3cm. It turned for about 10 seconds by itself, however a minute later the entire unit including the compressor heated up to the same temperature and the unit stopped turning. I had been lecturing that building thermal machines in the micro world makes no sense if you solve the equation of heat transfer, but I made the very mistake myself! It takes a real experience to know the stupidity you possess.

Knowledge when making decisions about building real products:
    (1) **The King's way of design:** Even if science and technology advance, we should not look down upon knowledge that our predecessors have found. These are the routes that everyone has to go through and working on them leads to success.
    (2) **Learning without thinking sheds no light:** To gain the real knowledge, you have to experience things for yourself without relying on verbal information from the teacher. My observations tell me that physical experience is the only way to get the knowledge sink into your brain.

The 17 Rules for an Engineer to Succeed

I think to decide means to set the undetermined (unknown) conditions and make a judgment. This means there is nothing to decide if things get settled logically, like solving a system of equations. We have to make decisions when our plans do not seem to proceed as expected. In contrast, when things go well, we can make decisions without much effort.

When, however, an assumption for a constraint changes, the circumstances change because of external disturbance, an organizational chain interferes, or when the health of your body or mind changes, the plan faces an obstacle and that is when we need to make a decision. I will explain this situation in 17 categories following the expansion of the mind in design:

[Setting the goal]	
Planning is gambling	(1) I want to make a big decision, but cannot meet all the requirements. Should I wait for the next chance? ➔ You'd better do it. There will be no next chance. Planning is gambling. Stop complaining and get started.
Better not do it if you need to postpone it	(2) Decide now, or do nothing and wait for time to solve it. For example, divorce, or bankruptcy. ➔ You'd better do it. Time does not solve a bad situation. Postponing it weakens the organization. Even when there is no progress, people will forget and that is why they say time solves it.
Information is always short	(3) There is not enough information to make the decision. Positive information keeps coming in but it is hard to believe. This is always the case with planning. ➔ Have the head of the organization make the decision. Only the president can decide whether to do it or not. He cannot blame others for bad information. Remember the assumptions and if any is wrong, correct it immediately.
Choose dream and will	(4) Whether to go with your dream or settle with reality? Insist on your desire? For example, you have something you want to buy, you want to go somewhere, or to choose a path with low possibility of success. ➔ You'd better do it. Doing it matches your mind better than not doing it. Unless you do what you want to, your brain does not get activated.
The first step	(5) Do not know what to do. The mind is unstable and cannot set the goal. ➔ Start with something small. It can be anything. Deciding what to do today matches your brain. A thousand-mile journey starts with the first step. For example, decide what to do next when you wake up in the morning. No novelty is required.

[Analyzing the problem]	
Virtual exercise on your own	(6) Your boss's boss suddenly asks for your opinion. Your brain is not ready for it. ➜ Always think ahead. Unless you have simulated it before, there is no way for you to know. Exercise virtually even if it may just go down the drain. This is what gives you the chance to get ahead. Your life in a corporation is shorter than you think.
You do not like who you do not like	(7) The constraints are all non-technical and you cannot find the solution. I can solve purely technical matters logically, but . . . ➜ Run away. You cannot do anything with the feeling of dislike. Putting up with it and advancing your career never meets the cost you put in. "If you hate the monk, you even hate his outfit."
[Selecting a solution]	
30% reason to a hunch	(8) Cannot quantitatively explain the decision. Follow what your mind says? ➜ Rely on your intuition. A hunch is 30% reasoning. It comes from your past studies. What you are thinking now is the smartest.
Laws of wisdom	(9) Follow the wise. Cannot come up with a good idea to make the decision. Take a bath? ➜ The wise one always thinks the same way. There is a rule for coming up with ideas. Consider the context and scenario, then a little thinking leads you to the solution.
Connection saves you	(10) Not enough information to select the solution. Cannot see where the breakthrough is. ➜ Use your connection (human network). Acquire and keep smart friends beforehand. Do not listen to others but have others listen to you. A consultant visualizes the customer's thoughts, and the customer learned what he wanted to say and is relieved.
Buy wisdom	(11) Can you believe in your own skills? Can you invent and develop just by yourself? ➜ Use others. No need to solve everything yourself. You can buy wisdom. Clarify your request without losing your target or problem.

[Deciding the detail]	
King's way of design	(12) Should I follow a previous solution, or challenge something new? ➔ Follow the example. A former example has its reasons in history. Years bring wisdom. The King's way of design. Former examples have concentrated experience to gain reliability.
Human resources	(13) Bothered too much by non-technical issues, especially vertical relation within the organization. Should you complain to the company? ➔ Ignore it. Do not think about it. Human resources determine everything. Avoiding unnecessary friction will turn out better for you. Fighting with others is negative work.
Irresponsible is the best	(14) Too busy and the mind is full. No slack for the body or brain. ➔ Better run away. Being irresponsible is good for your health. Do not live for the details. Even if you are not around, an organization grown so far will keep going.
[Executing the plan]	
Once decided, tackle it immediately	(15) The environment changed when we started. The original constraints have changed. ➔ Better not change it. Ignore external disturbances that were known. Exccute faster than the change. If things turns out badly, quit immediately.
Life without regrets	(16) Things do not always turn out right and how far should you place insurance? Should you negatively manage the risk? ➔ Better not change. Easier to run through thinking "Even if it suddenly falls, spend a life without regrets." Act positively and boldly. If things go wrong, call for help.
Analysis instead of regretting	(17) The overall project is not going well. It shouldn't have been undertaken. ➔ You are not the one to blame. Analyze the results instead of regretting it. If you regret, your mind goes down. If you can analyze yourself as if your mind was someone else's, that will be the best. If you drop your wallet and keep regretting it, you will lose twice as much. If you mind goes berserk, blame others. Blame your god.

- When the date to make the decision approaches, make it positive ((8), (12), (15), (16)), believing in what you have thought about. Of course, it is important to plan the worst case and prepare the counteraction ((6), (9), (10)). Along the same lines, when there are too many unknown constraints making the decision even harder, believe in yourself and leave it to your wits ((1), (2), (3), (4), (5)). You are the one thinking most about the plan, so believe in yourself and decide positively.
- Engineers, strangely enough, do not suffer greater stress when they decide on something purely technical. This is because for purely technical matters they can easily set the goal or define the problem, and most of the time can follow logic to arrive at the design solution. It is also due to the fact that many engineers like to think. Engineers certainly are not bored with logical discussions. To decide on something purely technical is not so burdensome as a big decision.
- The problem comes when even a purely technical difficulty cannot be overcome. The engineer is irritated, starts to doubt his ability, and is confronted with the negative decision of whether to quit the job or to stay. At this time, he builds up stress and grows gray hair ((11), (14), (17)). At these times, it is important to blame others and not yourself. In general, and this applies also to those that are not engineers, as long as we live in organizations, many of us worry about non-technical issues such as those concerning the organization or society ((7), (13)). As long as you belong to an organization, it is a waste of time blaming yourself when you are just one member of it. Of course, there are times when decisions that are negative for the organization are necessary, but you should make them with a clear mind. Decide quickly if you think your mind is losing it. Once you are really out of the line, you cannot even run away.

Our society is full of salaried businessmen and most of us think we have done our job if we complete our share of the work. Many of us never wonder "How do I relate to society through my job?" Maybe this is the source of the long recession in Japan.

After Snow Brand Dairy Product had the food poisoning incident, the ashamed executives of the company took on the morning habit of listening to recordings of claims from the customers before they started their day. The person who told me the story thought that Snow Brand would turn into a better company. "Really?" I said, "if that is what they are doing, they will repeat the same accident" and soon they had the false beef labeling incident. The company will now disappear.

Society now requires all employees from the president to the clerk think and act in their relation to society. It seems that companies that have had scandals have all shown the same symptom of disease from the root. Let's change the air in our companies. Let's change the culture. Otherwise, no matter what economic reforms we make, we will never escape the stagnation we are in now.

The other day, I purchased a pedometer. It only showed half the steps I took so I sent it back to the manufacturer, C Corporation. I then received a reply that said, "After testing the unit, it meets our corporate standards." I wonder what the inspector looks at. What does the president of the company do? Before we start talking about transferring production plants overseas or about the de-industrialization of Japan, does the president know that the ground is giving way under his feet? If the president is thinking along the same line as the inspector, maybe this company will follow in the footsteps of Snow Brand.

5

Real Decisions in Manufacturing

Chapters 5 and 6 contain articles by 40 different authors who documented their experiences in decision-making. The editor encouraged them to use illustrations in each article. These illustrations help the reader understand what the designer thought, how his mind process went, and the results of "decision-making."

Chapter 5 describes decisions made in the manufacturing field, Chapter 6 those about individuals and corporate directions. Section 5.1 introduces decisions in technology development, 5.2 those in technology management. The articles in each section are classified into the following categories:

(1) "Thinking" articles are creations that originated from past thinking or accepting new ideas.

(2) "Development" articles describe the successful commercialization of new products.

(3) "Practice" articles are those that may not be as big as development articles, but describe good ideas that led to new products.

(4) "Modification" articles describe creation by modifying existing technology.

(5) "Research" articles are pursuits of future dreams by challenging new undeveloped technology.

The classification of articles in Section 5.2 is as follows:

(1) "System Introduction" articles are about introducing new CAD/CAM technology to the business.

(2) "Technology Introduction" articles are about importing foreign technology.

(3) "Practical Solution" articles are about samples of starting from technology development to reach practical implementations.

The articles in this chapter are written with the author as the first person. By rewalking the paths the authors passed along, the reader will gain a good understanding of what decision-making means in manufacturing.

5.1   Decision-Making in Technology Development

5.1.1 Finding Out the Real Safety System at Mt Usu – Thinking 1

Event: We returned to the origin of developing a remote safety system in the hazardous area of Mt Usu. We were forced to overturn our concept of safety systems through our thorough checking.

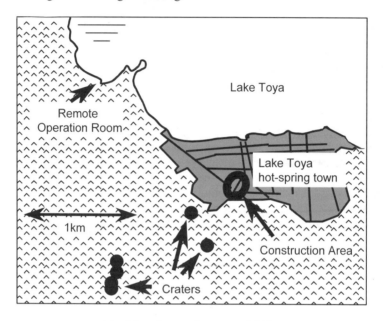

(a)   Map of erupted area near Mt Usu

(b) Looking at the construction area from the remote operation room

Figure 5.1   Remote construction site near Mt Usu

Course:

March: On March 31, 2000, Mt Usu erupted after 23 years dormant. New craters opened near the hot-spring town of Lake Toya.

April: Volcanic mudflow from Mt Usu drifted into the town. We had to save the town from disaster.

May: Started "Mt Usu Unmanned Construction Project."

The construction work took place at the close range of 300m to 700m from the erupted craters. We sent unmanned remote control machines into the hazardous area. Figure 5.1 shows the layout of the construction site. Removing the mudflow and securing channels to guide it were the urgent tasks.

June: The control was frequently lost for minutes on end due to control signals getting mixed up.

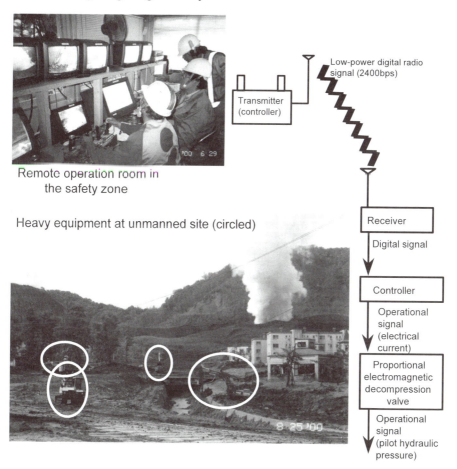

Remote operation room in the safety zone

Heavy equipment at unmanned site (circled)

Low-power digital radio signal (2400bps)

Transmitter (controller)

Receiver

Digital signal

Controller

Operational signal (electrical current)

Proportional electromagnetic decompression valve

Operational signal (pilot hydraulic pressure)

Figure 5.2   Wireless system for remote machine control

Motive: Failure of the unmanned machines could have destroyed the entire construction work, and therefore, we had to come up with appropriate measures as soon as possible to minimize the harms.

Action: We investigated the unmanned machine radio system extensively, searching for the cause of the problem and the solution. The construction work was a 7-days-a-week operation, and the sites were closed for the night. These factors made it impossible to perform operational tests on the unmanned machines. The radio system had been in the field for over 6 years without any problems, so we looked into the circuit and software, going back to the starting point of the development.

Investigation: Figure 5.2 shows the radio system configuration of the unmanned machine controller. An operator at the remote control station led the unmanned machine, watching the images coming from the machine.

The unmanned machine had a fail-safe system to prevent faulty operations in case of error in the low-power digital radio control signal; the engine automatically shut down in this case. Once the engine was shutdown, the system required normal signal reception for an extensive time period to resume its operation. These robust fail-safe requirements caused the machine to stall.

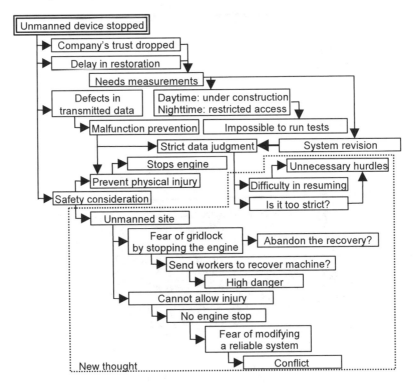

Figure 5.3 Expansion of thoughts diagram of what entered the mind when the unmanned machine control system failed

Hesitation: We first asked ourselves about the feasibility of applying an ordinary fail-safe system to the unmanned operation. Figure 5.3 indicates a conflict at the time when the question was brought up.

"Modifying an existing system, which had reliably been in operation for an extensive period of time, may have been the decision. We must face what is happening in front of our eyes, and we need a way to resolve the problem."

Figure 5.4 shows the fundamental difference between an ordinary remote-control system and the unmanned remote-control system used in Mt Usu.

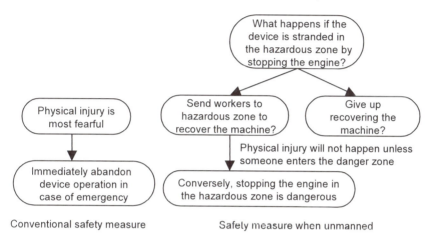

Figure 5.4 Conventional safety and safety when unmanned

Decision: We reversed the regular concept of safety measures and developed a set of new rules for the unmanned situations. Figure 5.5 shows the modification.

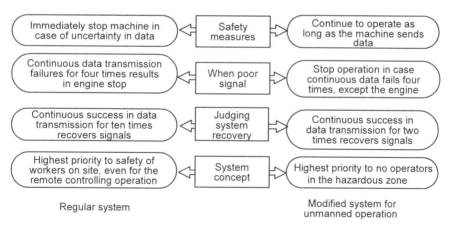

Figure 5.5 Conventional safety measures and new measures

Results: The time for system recovery from machine failure caused by a poor signal was reduced significantly.

Lessons:
- In the regular sense, safety requires stopping the machine in case something goes wrong. In the regular world, it is hard to think of an operator risking his life to restart a stopped machine.
- Know that your mind is always surrounded by blocks!
- Let your mind see what is in front of your eyes. Do not let old concepts fool you.
- Sometimes you need to overturn existing concepts.
- It is important to view and check the problem from the starting point, if the situation differs from past situations.

Sequel: The urgent construction work progressed on schedule, and peaceful days returned to Lake Toya's town of hot springs.

5.1.2 Designed a Pressure Sensor Exposed to Severe Conditions – Thinking 2

Event: I received a request to develop a pressure sensor to measure the pressure in a container which was subject to severe environment. The task was considered generally impossible, so I applied the expansion of thoughts diagram in the hope of achieving a breakthrough. The diagram led me to a new idea that enabled designing and putting the sensor into practical use.

Background: Containers used in that industry are often exposed to instantaneous high rises in temperature and pressure. Keeping track of sudden pressure spikes in a container is crucial to maintaining the safety of industrial processes. Ways of measuring instantaneous pressure rise existed, however there were none for measuring the negative peaks that follow. We needed to measure such pressure values.

Requirements: The container, initially at ambient temperature and pressure, goes through instantaneous high rises in temperature and pressure followed by negative pressure. We needed to accurately measure the negative pressure. The negative pressure sensor system had to meet the requirement in the following paragraph.

Figure 5.6 indicates the pressure time history of the container. Starting from ambient temperature and pressure, it was exposed to high temperature and pressure for approximately 20milliseconds (2,000°C maximum temperature and 400MPa maximum pressure), and then to rapid temperature and pressure decrease. The plan was to measure the negative pressure with an accuracy of 0.001MPa.

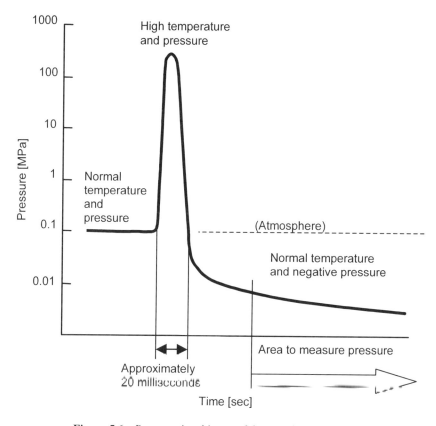

Figure 5.6   Pressure time history of the container pressure

Solution: The required measurement accuracy was 5 digits smaller than the pressure exposure of the sensor and there was no such pressure sensor in the market. Figure 5.7 shows the expansion of thoughts diagram I generated in devising the breakthrough.

By expanding my thoughts, I separated the functional requirements into one for the valve and one for the sensor. This division made the accurate measurement of the pressure possible. I then designed the valve and sensor separately to meet their functional requirements.

Figure 5.8 shows the operational requirements for the valve at time steps, I, II, and III. The pressure was low at stages I and III whereas the valve only opened at stage III. I then listed the conditions in a search for the structure that realized the operational requirements, and reached the idea of material that vanished when exposed to high temperature and pressure. Figure 5.9 shows the expansion of thoughts to reach the valve that met the requirements.

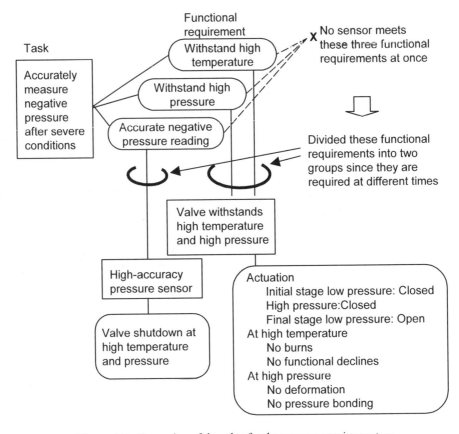

Figure 5.7   Expansion of thoughts for the pressure sensing system

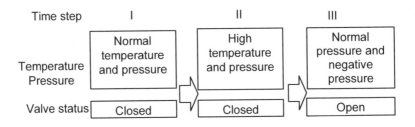

Figure 5.8   Summary of operational functions of the valve

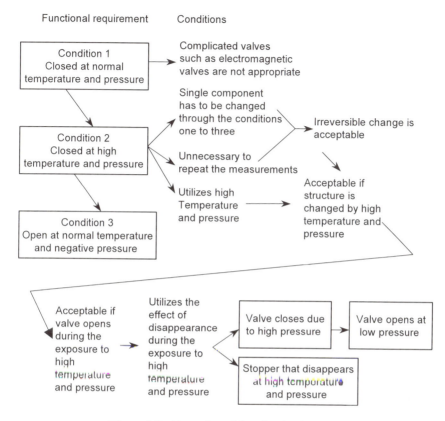

Figure 5.9   Expansion of thoughts for the valve

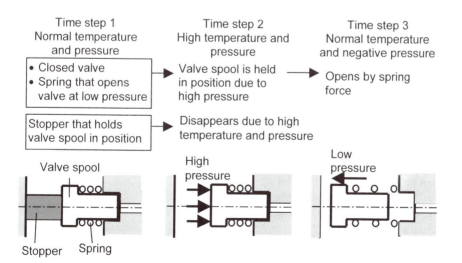

Figure 5.10   Concept of the valve

Figure 5.10 shows the concept of how the valve was to be structured. The stopper, at the beginning, kept the valve closed in the first stage. When the structure was exposed to high temperature and pressure the stopper vanished, however the valve stayed closed due to the surrounding high pressure. When the pressure dropped to low again, the spring pushed the valve spool out and let the gas the reach the sensor. Figure 5.11 shows the actual structure of the valve.

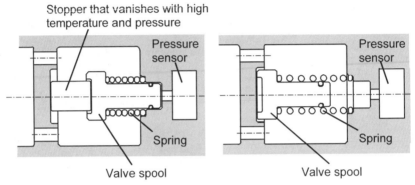

Figure 5.11   Structure of the valve

Results: We ran some material tests, calculated strengths, pressure, and spring responses and built a prototype. The overall test proved the validity of my concept and we successfully measured the container pressure transition after the event of exposure to severe high temperature and pressure.

Knowledge: Even though a problem may seem to have no solution, the expansion of thoughts diagram clarifies constraints and helps in reaching the target.

In this particular case, we reached the breakthrough by practicing "separation of time" and "introducing a third material," a couple of techniques that the "principle of creative design" teaches as ways of expanding thoughts.

5.1.3 Reduced the Weight of an Automobile Compressor – Thinking 3

Event: We successfully reduced the weight of the compressor for an automobile air-condition by removing the balance weight for the crankshaft.

Background: An air-conditioner is standard equipment for recent automobiles and is installed on the assembly line in a factory. This story goes back to the 1980s when an air-conditioner was not standard and was installed at a dealership instead of at the factory. Because of the unit's weight, air-conditioner installation largely increases the total vehicle weight. The consideration of the added weight led to legislation stipulating that an air-conditioner must be under 20kg. The air-conditioner manufactures were forced to make every effort to meet this standard. The compressor, which is a major part in the air-conditioning system, was no exception, and its weight limit of 5kg dropped to 4kg. Air-conditioner manufactures started to tear their hair out to meet the new requirement.

We first tried to solve the problem by taking some mass off from the crankcase by changing the piston bore size and stroke, but the modification was not enough to reach the target.

Solution: Generally, the crankshaft contains a balance weight, and we removed it to meet the targeted weight reduction. Contrary to our concerns about excessive vibration, the engine operated smoothly and there was no sign of severe vibration under operation at 9,000 rpm, *i.e.*, 1.5 times the normal rotation speed.

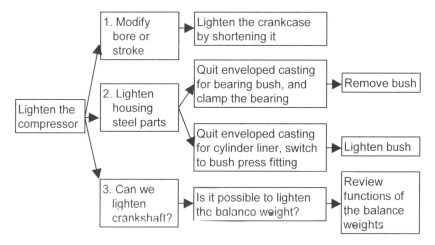

Figure 5.12   Expansion of thoughts for lightening the compressor

Mind Process: Figure 5.12 shows the steps of the mind process. The purpose was to "by any means, lighten the compressor."

(1) Change the (piston diameter) x (stroke) and lower the crankcase height.

(2) Lighten the steel parts attached to the housing. The original design had a steel bush cast into where the bearing was press-fitted. We got rid of this steel bush by clamping the bearing in the axial direction. Also the aluminum housing had a thick cast-iron bush for the cylinder liner; in its place we press-fitted a thin sintered steel bush.

(3) I then realized that the crankshaft was also made of steel, and wondered if I could reduce its weight. The first thing that caught our eyes was the balance weight. Was there any way to lighten it? My thought went to the purpose of the balance weight. Common sense (without reason) says that an inline two-cylinder compressor requires balance weights to counter the force of inertia created by the piston, gudgeon pin, con-rod, and other parts in reciprocal motion.

More precisely, the phases of the two pistons differ by 180 degrees, to counterbalance inertia in the vertical direction. The horizontal distance between the two pistons, however, creates inertia (moment of a couple), as Figure 5.13 shows. This moment causes vibration.

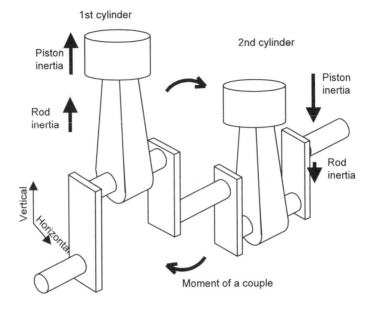

Figure 5.13   Pistons generate a moment of a couple

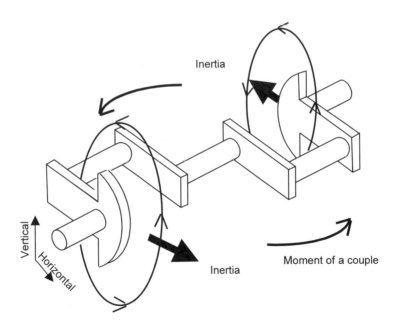

Figure 5.14   Moment of a couple from the balance weights

Despite the concept, however, the actual balance in place only amounted to about half of the inertia from the pistons' motion. The reason is as follows. The inertia from each piston stays in the plane spread by the crank axis and the cylinder axis. Each balance weight, on the other hand, rotates about the crankshaft axis as Figure 5.14 shows. The balance weight, therefore, does not really counter the piston inertia. The two balance weights generate a moment of a couple in the horizontal plane, and the moment is bigger with heavier weights. The balance weights are a tradeoff between reducing the moment of a couple in the vertical plane and increasing that in the horizontal plane. I then concluded that if the compressor is fixed to the engine rigidly, the balance weight is not really required.

(4) I followed the logic above and reached the idea of getting rid of the balance weight. Figure 5.15 shows this mind process, and Figure 5.16 the cross-section of the actual compressor.

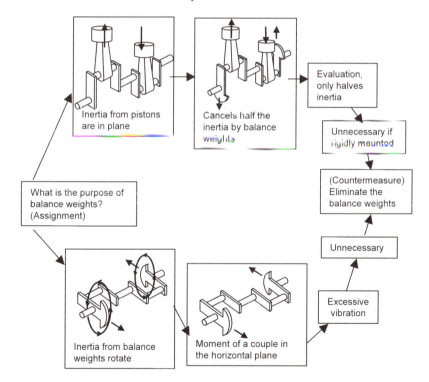

Figure 5.15   Mind process about the balance weight

Knowledge:
(1) Reciprocating machines with in-line cylinders cannot cancel the moment of a couple even with balance weights.
(2) Parts, for historic reasons, are sometimes put in place without thorough thinking. We have to discover their real purposes.

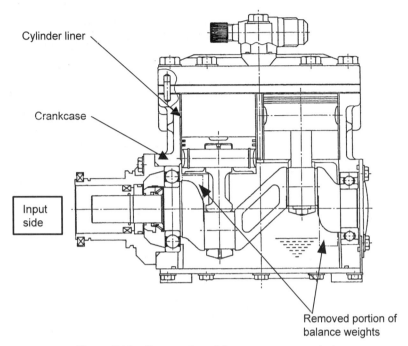

Figure 5.16   Cross section of the new compressor design

(3) We sometimes have a mind block such as, "Removing the balance weight makes the vibration larger." We sometimes have to go to the very root of what we understand as "common sense."

Side notes: In addition to this redesign, I changed the solid crankpin arm from solid to one with an H cross-section, and opened a hole in the far side from the input. When pushed, people make things happen by any means they can.

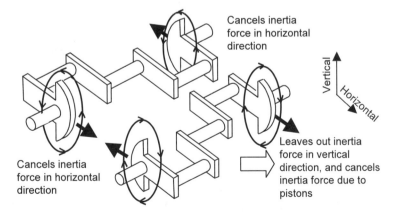

Figure 5.17   Removing horizontal inertia with two shafts

5.1.4 Automated the Narita Express Car Junction Hood – Thinking 4

Event: A project was started for developing an automatic junction hood for Narita Express (NEX). The development was in response to a customer request. At the time, we were translating the book *Principle of Design* by N.P. Suh and we applied axiomatic design to come up with an automated car junction hood with a new structure. Our development was the first for such an automated hood in Japan.

Background: Our customer requested a development project for automating the junction hood. NEX was planned to connect major hub stations of the Tokyo metropolitan area and Narita International Airport. NEX started its operations in March 1991 and the cars departing from and arriving at Shinjyuku and Ikebukuro were joined with and separated from those for Yokohama. The joining and separation took place in Tokyo station in the underground level. These operations went on next to the platform and required the safe connection and release of the junction hood. The conventional junction hood operation took a long time and a number of operators. Automating this operation would reduce the connection/separation time and the workforce.

Functional requirements and constraints: Figure 5.18 shows the functional requirements and the constraints.

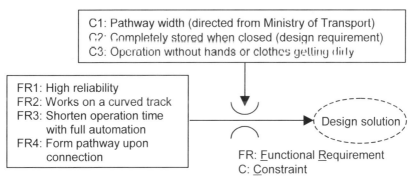

Figure 5.18   Functional requirements and constraints for automatic junction hood

Mind process and hesitations:

Figure 5.18 shows the functional requirements and the constraints.

(1) Initial design concept and customer evaluation: The initial idea modified the existing bellows junction hood by adding an automated mechanism to it; this was a solution that axiomatic design completely rejects.

This initial design automates the existing bellows junction hood; for making the connection, it had to (i) Push out and hold the hood base in place, (ii) Extend the hood frame, (iii) Position the hood frame, (iv) Fix the hood frame, (v) Release the extending mechanism, and (vi) Store the extending mechanism. There were too many steps, which made the solution unnecessarily complex. The customer stated that the complexity of the mechanism jeopardized the reliability (and I thought so, too).

(2) Second design concept in the functional region: When we presented the first design concept to the customer, we learned that one of the customer conditions was that the new hood design did not need to join with the existing hood. We could completely forget about the existing hardware, and we could start our design from scratch in the functional region. We then applied axiomatic design and expanded the functional requirements (FR) and constraints (C) as shown in Figure 5.18. Then we expanded our thoughts as Figure 5.19 shows.

To satisfy FR1(high reliability), we wanted to simplify the mechanism with fewer steps. We then came up with the idea of applying the western-style rubber hood (Figure 5.20 (b)) instead of the Japanese-style bellows hood (Figure 5.20 (a)).

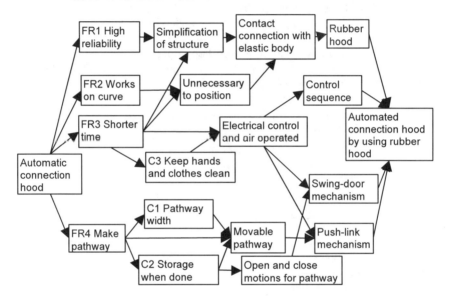

Figure 5.19   Expansion of thoughts for automatic junction hood

A rubber hood forms the pathway by simply pressing the rubber tubes, attached to the end of the two cars to be connected, against each other. It does not require a mechanism to join and hold the hood, and its structure is simple. It has, however, the shortcoming of opening a gap on sharp curves. The rubber hood is widely used in the western world, maybe because of the high sense of responsibility of the user.

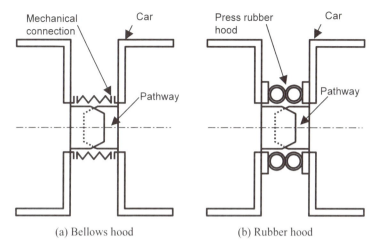

(a) Bellows hood   (b) Rubber hood

Figure 5.20   Bellows hood and rubber hood

The automatic junction hood with these rubber tubes only required a single step of (i) extend the hood base and hold it (press the rubber). The mechanism was greatly simplified, meaning a large increase in reliability. The positioning that the bellows hood required was no longer necessary and FR2 (works on curve) was also satisfied. We then compiled the second design concept of the automatic junction hood.

(3) Setting the design solution: Figure 5.21 compares the two design solutions. The second concept of rubber hood scored higher over the first concept in terms of FR1 (high reliability), FR2 (works on curve), and FR3 (short time), however, in terms of FR4 (make pathway), it loses because the pathway opens a gap when it is on a sharp curve.

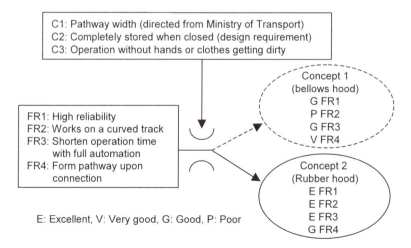

Figure 5.21   Comparing the design concepts (tradeoff)

We then evaluated the conditions of operation. The purpose of the pathway is for the attendants and emergency evacuation. All seats on the train are reserved and the passengers do not walk through the pathway. In addition, the NEX lines do not have any sharp curves that would open gaps in the hood.

We thus concluded that FR4 (make pathway) was not crucial and winning in terms of FR1 (high reliability), FR2 (works on curve), and FR3 (short time) had higher priority (tradeoff).

Realization and Results:
(1) Second design concept and customer evaluation: The customer liked the second idea and it was adopted for use in NEX trains.
(2) Development and production: Despite the tight schedule, we developed the automatic junction hood (Figure 5.22) to meet the scheduled opening of NEX lines. After it started operation, it had some minor troubles, but is now safely and stably in operation.

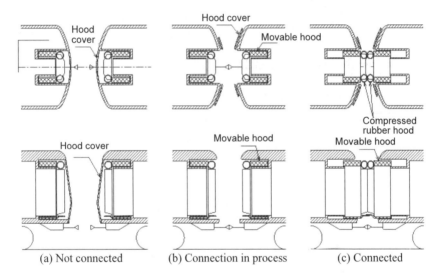

(a) Not connected   (b) Connection in process   (c) Connected

Figure 5.22   Conceptual sketch of how the automatic hood works

Reflections: This automatic junction hood, however, targeted only one car type and we could not apply the same design to other car types. The users, concerned with safety issues, did not like the gaps opening on sharp turns. Other cars have the bellows-type Concept 1 automatic junctions hood that do not work over curves, have complex structures, but have no gaps when the car is running over curves.

Evaluation: Properly stating the functional requirements and constraints allowed us to reach a solution for practical use. I wrote about this case as a successful accomplishment in my engineering exam and I passed.

Knowledge:
(1) When adding a new function to an existing design, always organize the functional requirements and constraints in the functional region (This is what Professor Suh claims in his axiomatic design theory).
(2) In addition to the customer functional requirements and constraints, add your own product plans.

Targeting the 2002 FIFA World Cup soccer tournament, NEX added 2 new arrangements, taking the total number to 23. The automatic junction hood is still in use at the Tokyo station underground level, where you can actually see it in operation.

5.1.5 Developed a Telescopic Arm Clamshell Digger – Development 1

Event: We were the last to enter a market where five other leading companies had already been in severe competition and swept away most of the market.

Background: Our company always wanted to enter the market for work on the underground foundation work with urban-type construction projects. However, we had no telescopic arm clamshell digger in our product line and needed to develop one. Five other companies were already in the market, each with its own share, and our development had to be a breakthrough product that could take sales away from predecessors.

Did we have any chance with such a late entry into the game?

Outline of the machine: The telescopic arm clamshell digger has a tube arm that extends by three stages with a clamshell-like digger at the tip. This arm is installed on the base structure of a backhoe. Its main use is to excavate soil from holes that are about 20m deep and dump it in a truck. Urban-type underground constructions often require this operation (Figure 5.23).

Design concerns and constraints: We surveyed the users about the machines in the market and found the following three complaints:
(1) The wire that made the three-stage telescopic tube extend and contract would sometimes break, dropping the clamshell.
(2) This wire had a short life that could be as brief as 1 month, and exchanging it was expensive and time-consuming.
(3) The machine had a long and heavy telescopic tube, and to balance out the whole machine, the clamshell was small. The users thought they had no choice here due to the structure of the machine.

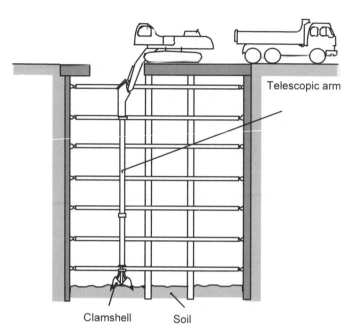

Figure 5.23   Using a telescopic clamshell for underground construction

We, being the last to enter the market, and also having to succeed, had the following development concerns in addition to having to overcome the above problems.

(4) The product had to be strong and at the same time the strength had to be appealing to the user. The sales force would have then confidence when they went round the customers to explain the product. This would be helpful in introducing a new product in the market when the sales force was not so large and the product had no sales records.

(5) A new entry in the market would drop the price and we needed enough margin above the product cost to make a profit.

Base structure: The important point at this stage was to select the best base structure among all possible options. A mistake here would cost a large loss later. We discussed the base structure with the following expansion of thoughts and set the final design.

(1) (Step 1) Expand the above five concerns to the functional elements and further to solutions. Usually, this step would have a number of different ways of expanding the thoughts, making it hard to reach a single solution. We came up with five different solution plans. Figure 5.24 shows the expansion of thoughts diagram and Figure 5.25 the five different solutions we evaluated.

(2) (Step 2) We evaluated the level of accomplishment for the five different solutions and conducted an overall evaluation (Figure 5.26).

(3) (Step 3) From the five options, we selected the wire-plus-hydraulic-cylinder type (Figure 5.25 (5)) as the base structure because it addressed more

concerns than the others. The single hydraulic cylinder on the tip arm extended to push the tip arm out from the middle arm and at the same time the two pulleys and wires on the middle arm pushed the middle arm out from the base arm.

(4) (Step 4) The option we selected did not solve all the concerns. It left the issue of the clamshell dropping when the wire broke. We then added a level bar as shown in Figure 5.27 to make the final design. When one of the wires was about to break, it extended and tilted the level bar. By monitoring the level bar angle, the design would allow the detection of wires that were almost broken.

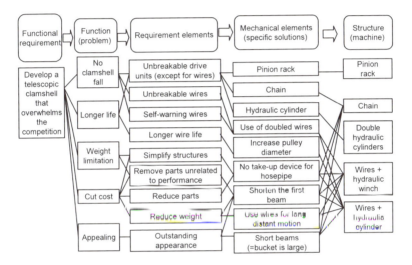

Figure 5.24   (Step 1) Expansion of thoughts diagram for the base structure

Product completion and commercialization: We had to face the problem of overweight during the design and prototype phase; however, through believing the decision was the best and with considerable efforts, we completed the product's development and introduced it on the market.

Knowledge:

- When solving more than one problem (in this case five), we should plan a solution that solves multiple problems at the same time. Tackling one problem at a time would lead to a system that is out of balance and can cause other problems.
- If unable to find a solution that solves all problems, pick the best and concentrate on solving the shortcoming afterwards.
- Late entry into an established market is not necessarily something we have to worry about. The leaders, once they have established themselves, tend to stick to the same base solution and make small modifications. There may be other unexplored and attractive ways to go. The newcomer may very well uncover a whole heap of treasure.

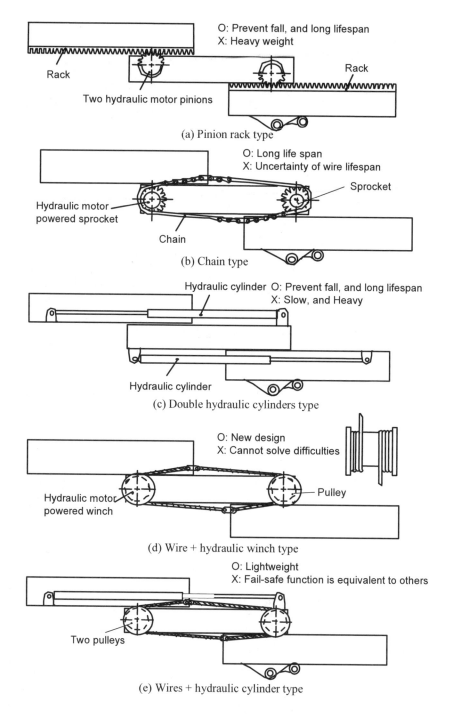

(a) Pinion rack type

(b) Chain type

(c) Double hydraulic cylinders type

(d) Wire + hydraulic winch type

(e) Wires + hydraulic cylinder type

Figure 5.25   Five structures that we evaluated

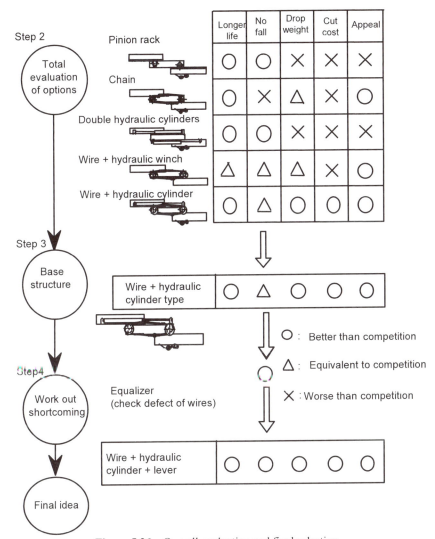

Figure 5.26   Overall evaluation and final selection

Sequel:
(1) The product we developed ended up with an overwhelming competitiveness over other products in the market and it turned into the primary product for underground construction work.
(2) When we compared a competitor product (Figure 5.28) with our development (Figure 5.29), the strokes were the same, however our product had a simpler structure and weighed less. The sales force pitched to the user that ours had "a larger clamshell for faster construction." Our development allowed using a larger pulley and hence imposed less bending on the wires. The wire life extended about three times and, thanks to the balance lever, we never heard of a clamshell dropping accident.

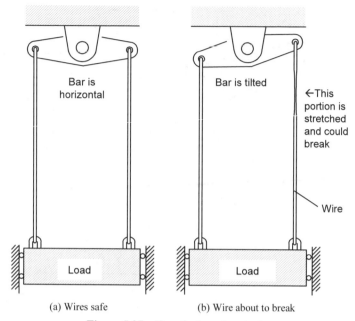

(a) Wires safe   (b) Wire about to break

Figure 5.27   How the lever bar works

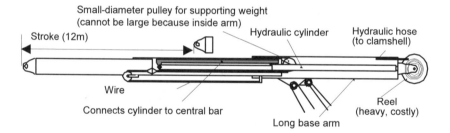

Figure 5.28   Competitor's arm

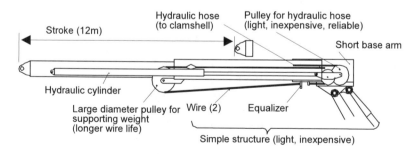

Figure 5.29   Arm that I developed

5.1.6 Developed an Automatic Segment Construction Robot for Sealed Tunnels – Development 2

Event: As a tunnel machine bores a tunnel, it places ring-shaped tunnel support parts called segments which are structured to resist the soil pressure. The segments add strength to the inside wall of machine-drilled holes (Figure 5.30). Some of the most recent tunnel-boring machines drill large holes of over 10 meters diameter and the segment placing proceeds automatically. This article describes a project in which the initial development strategy and concept turned and twisted in the course of its process and eventually succeeded in developing an automated high-speed construction machine.

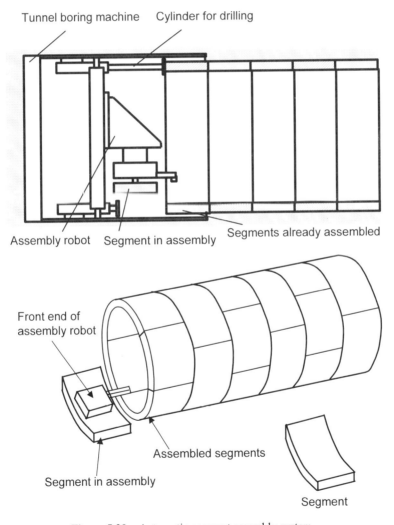

Figure 5.30   Automatic segment assembly system

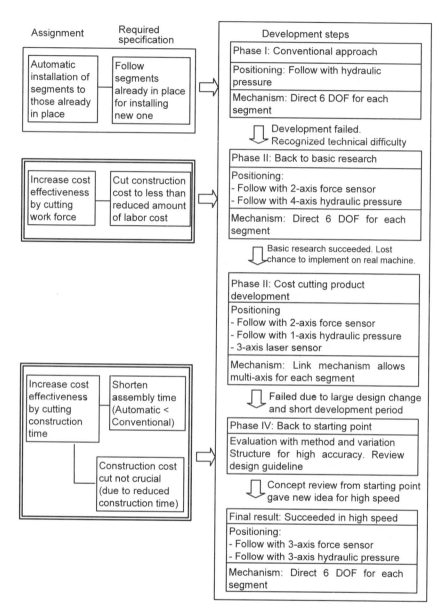

Figure 5.31   Expansion of thoughts and systems at each phase

Course: Figure 5.31 shows how the project strategy and concept changed.

The project aimed at (1) safety and (2) saving power (manpower). The technical challenge was to install a new segment quickly and automatically onto segments already placed. The relative position of the assembly robot against the segments already installed kept changing, and thus sensing and positioning with six degrees of freedom were necessary. During the positioning, alignment interference with aligning in one direction causing misalignment in others occurred easily and we needed to devise a solution that quickly finished the process.

(1) **Phase I** (the first four years were the originally scheduled project length)
   (i)   **Background:** At the time, we were trying to win the bid for constructing the Cross Tokyo Bay Road and we needed to prove the feasibility of our technology. We started the project as a joint venture with M Construction.
   (ii) **Strategy:** Involving a new technology, the project was recognized as a basic research; however, just like a regular sales project it had to be profitable.
   (iii) **Course:** Targeted establishing the basic technology through in-plant basic tests; however, the tests took longer than scheduled and the machine malfunctioned on the demonstration day. We could not produce automatic results during the field tests, either. The short development time, our limited experience in automatic assembly, and steel segments with poor dimensional accuracy for small diameters all gave us problems in positioning the segment pieces.

(2) **Phase II** (4th to 5th years, marking our restart from the basic research)
   (i)   **Background:** Letting the division lead the research project led to a failure and we changed our strategy. Judging that requiring a short-term profit on an R&D project would delay results, we gave the project to a research center.
   (ii) **Strategy:** Recognize the technological difficulty of the project and prioritize establishing the technology.
   (iii) **Course:** Tried the development concept of "faster speed with following control." Control method was almost complete for a single-ring assembly test in the lab. Some concerns remained for putting the system into production. The demonstration for M Construction went well; however, we lost the chance to put it on a real machine.

(3) **Phase III** (5th to 8th years, product development with cost cutting)
   (i)   **Background:** Lost project with M Construction, and thus also the Cross Tokyo Bay Road opportunity. Our division, however, had high hopes and set out for in-house development of a general machine. Had to significantly cut the production cost to meet the cost-saving expectation from manpower reduction.
   (ii) **Strategy:** Cut the production cost for better cost performance.
   (iii) **Course:** Switched from an orthogonal slide driving mechanism to a link mechanism to cut the product cost (Figure 5.32). The decision was pushed from necessity; however, the question remained whether it was the right one in terms of engineering. This modification complicated the engineering for following control (Figure 5.33) and the research almost went almost all the way back to the starting point. Constraints of the design forced us to use additional laser sensing for positioning with following control. Additionally, management was told that the basic research had been complete, and we were pushed to turn in the results quickly.

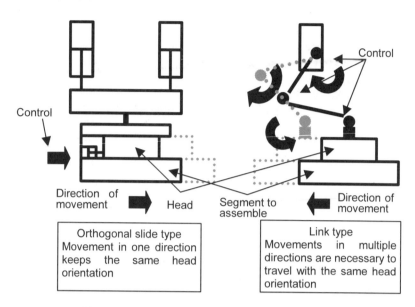

Figure 5.32   Orthogonal slide and link drive mechanisms

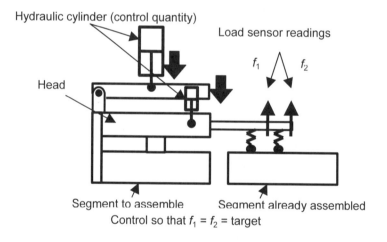

Figure 5.33   Example of following control with a load sensor

As a result, a structural change by applying a link mechanism turned into the main development issue, requirements for positioning accuracy did not reflect back to the design and the system had latent problems inside.

A number of problems surfaced in the total testing. The system seemed to have cleared the target performance, but when it came to the final evaluation it failed the demonstration. The going back and forth was repeated several times.

Each time a problem surfaced, the level of the system went up a notch, except for the positioning reliability. The developer gradually lost the respect paid to him. The manager in charge was upset with the fact that the orthogonal system was going well but they had to face new problems for switching to the link system. We knew in our hearts that the problem was in the fundamental structure, but we were at the stage of total testing and had to quickly fix minor problems. There were no time to go back to the very basics. After failing several evaluation tests, the link system was rejected and the project had to go back to the starting point.

(2) **Phase IV** (8th to 9th years, reviewed technology and concept)

(i) **Background:** With several failures repeated as stated above, the division manager stopped hurrying us, but gave us one last chance.

(ii) **Strategy:** Find a solution and the best one without being rushed.

(iii) **Course:** It turned out that this strategy worked well. We were able to tackle the problem squarely and discussions about the instrumentation and variation had clarified some problems. Through thorough discussion of a system to accomplish the required positioning accuracy, we were able to reach a solution that could both meet the manpower reduction goals and achieve high speed.

The new concept was based on following control and the mechanism was made for following (when switching axes, those already set would remain fixed). The concept was easily arranged for high speed and would make a great contribution to shortening the overall tunnel digging projects. M Construction ordered the system and we successfully developed the high-speed automatic segment construction robot.

In this phase, user requests and other constraints often tried to push the design to contradict our new concept. We only accepted those that coexisted with our concept.

We successfully applied the automatic segment construction robot for completing the construction. We saw an increase in orders for tunnel machines construction work.

Knowledge: Development strategies are important for the success of development projects.

(1) Thoroughly understand the technical difficulty.

(2) Technology is built in steps, one at a time, and development takes time. First concentrate on solving significant concerns and lower the risk in the early stages.

(3) In addition to lowering the cost, we can also modify the development concept and raise the technical value. We should concentrate on promising technology with enough margins and seek solutions for the development task. We should avoid bringing in unnecessary problems that complicate the goal.

5.1.7 Developed a System for Preventing Mobile Crane Overturn – Development 3

Event: We developed a safety system that automatically shuts the rotary motion off to prevent mobile cranes from tipping over. We applied the development to further shut off any abnormal movement, and the system was installed on commercial systems.

Background: Mobile cranes are mandated to have the safety system called overload prevention to avoid tipping over or slamming into buildings. The overload prevention stops such movements as hoisting loads, lowering the boom, and extending the boom upon detecting dangerous situations. The rotary motion, however, only generates an alert without stopping because stopping a rotary movement can cause the hanging object to start swinging. The technology to stop the rotary motion without causing the load swinging was not available.

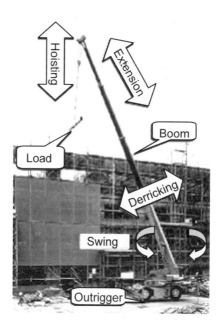

Figure 5.34   Mobile crane operation

A mobile crane with wheels for its transportation, *e.g.*, a truck crane or a rough terrain crane, are equipped with outriggers (Figure 5.34, Figure 5.35). The outriggers are stored under the body during transportation, and extend out for the crane operation to stabilize the crane body. The outriggers are located at left and right, front and back, to stabilize the crane and increase its capacity for hoisting loads. Conventionally, fully extending and landing the four outriggers stabilizes the crane and rotating the boom does not affect the crane capacity. Figure 5.36 shows the diagram for the setting the load capacity. The net loading (crane operation capability) is based on the boom length and radius of operation (horizontal distance from rotary motion center to load), and is independent of the angle of rotary motion (direction of boom). In this case, the critical movements are: lifting, which increases the load (when the load is lifted off the the ground), lowering and extending the boom, which increase the radius of operation. Rotary motion, therefore, does not require automatic shutoff.

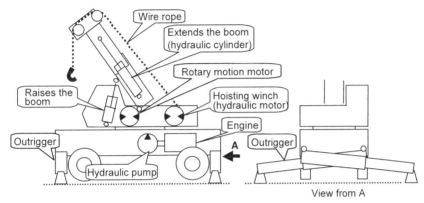

Figure 5.35 Hydraulic driving system of a mobile crane

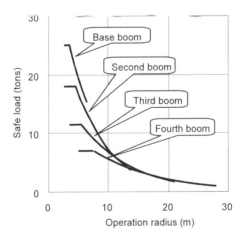

Figure 5.36 Conventional load capacity display

Circumstances that surround mobile cranes, however, have changed. When in a residential area, the crane can only use half of the full road width. The outriggers, therefore, can fully extend on the sidewalk side (=main construction site), but not on the road side. Crane manufacturers have developed ways to partially extend outriggers. In this case, however, if the crane has a load fully extended to the limit, rotating the boom to the road side would overturn the crane because the outriggers are not fully extended on that side. Figure 5.37 shows how the safe load depends on the crane operation radius and rotary motion angle. The safe load decreases at the front side of the figure, in this case, the road side. Under these circumstances, an automatic shutoff function for the rotary motion as well is necessary.

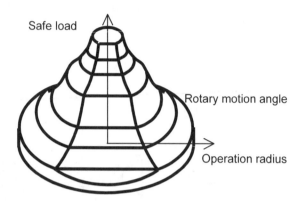

Figure 5.37   Maximum allowable load with outrigger extended

Motivation: We started our technology development in 1988. Those in the team had already been discussing the need for such a safety system to automatically arrest the rotary motion. Japan was in its bubble economy in 1988, and a number of overturn accidents had been reported from construction sites. NHK broadcast a documentary about overturn accidents and crane safety issues, and it drew much attention from the public. Company A, which I was working for, were proud of being the top crane manufacturer, decided to make an effort to develop automatic shutoff technology for the rotary motion.

Investigation: We had already been conducting research into automating the rotary motion before we started the development of automatically shutting off the rotary motion. Although the original purpose was to develop the auto rotary motion shutoff safety device, we assumed that any automatic operation would be possible if the most difficult automatic control of rotary motion was achieved, and therefore we focused our research on the study of automatic operation of the rotary motion from actuation to the end.

   We modeled the rotary motion with pendulum motion on a plane. Figure 5.38 shows that a pendulum moving at a constant speed can be stopped without further swinging if we apply constant deceleration that matches the natural period of the pendulum to its pivot point. Figure 5.39 shows the crane rotational velocity, load velocity and swing angle when the cable length was 19.6m and initial velocity was 0.052rad/s. The same behavior applies to acceleration and this control had already been widely known. In the research field of automatic operation, the on-off types bang-bang torque control in two deceleration stages was also well known; however, response delay in the rotation driving system, non-linearity, and deformation of structures added complications, and we could not find a good answer to minimize the swing after the load had stopped. The most significant disturbance came from the swing of the load at the start of the stopping control, that largely affects the swing during the control. We, at the time, had concluded that we could not get good control without sensing the amount of swing. We had also tried developing a swing sensor, but had not reached a good solution for practical use. The research of automatic rotation was at a stall.

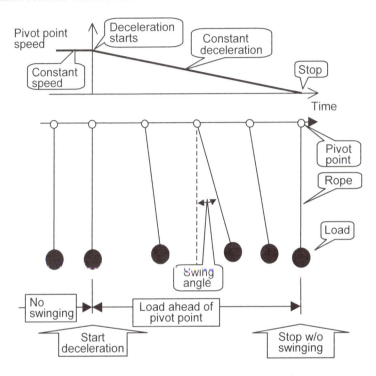

Figure 5.38   Swinging with constant deceleration

Operating a crane involves four basic movements of winding the load up/down, raising and lowering the boom, extending the boom, and rotating it. A mobile crane is driven with hydraulics and as the engine turns the hydraulic pressure operates the wind up/down winch, the boom raise/lower cylinder, the extension cylinder, and the hydraulic motor for rotation (Figure 5.35). The operator manipulates levers to control the flow and direction of oil pressure using control valves. The control valves allow adjustment of the speed of each movement. The winding up and down operation is only in the vertical direction and thus it is relatively easy. The extension operation is not usually performed with a load hanging. The raising and lowering of the boom are slow operations and proceed without causing any swinging of the load. The most difficult operation is the boom rotation and the centrifugal force in the radial direction and force in the direction of rotation cause complex swinging of the load. The situation worsens if the boom is extended far and the wire cable is rolled out a long way. Furthermore, boom rotation moves the load and crane in the lateral direction to greatly change the positional relation of the work area and load. At the time, rotation was considered as work to be carried out only by skilled operators.

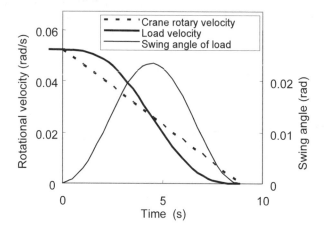

Figure 5.39   Crane velocity, load velocity, and swing angle

Hesitation: We ran some computer simulations and confirmed that the load swinging motion at the beginning of stop control largely affects the following swing motion and even makes the swinging increase in some cases. Naturally, faster rotation produces greater swinging. It seemed that a simple pendulum in a plane was an oversimplified model. Research papers and operational experience suggested that we needed to operate the boom lift/lower lever to add radial direction control to completely stop the load from swinging when we wanted to stop the rotation. So, did we really have to develop a swing motion sensor to accomplish our goal?

The source of my hesitation was that the real cause of overturn accidents from rotating the boom were unknown. We knew that the accidents happened when some outriggers were only partially extended, but, that was all we knew. How was the operator manipulating the machine then? Did he not know the outrigger settings then? If the cause was a rough rotation by the operator, we could not accomplish our goal without a swing sensor. Proceeding with our development without knowing the real cause would produce meaningless results.

Assumption: We then adopted the following hypotheses to explain the overturn accidents. If these hypotheses were correct, our target system would meet our goal (Figure 5.40).

(1) The operator knows that not all outriggers are fully extended out.

(2) The operator is at a sufficient skill level and always operates the crane carefully. His speed of rotation is slow and the amount of swing is small.

(3) The operator is always attentive to the surrounding work area. He is extremely cautious when he is rotating the boom because the work area and crane are about to make a big change in their relation. His attention is all on

the rotation operation and he forgets about the partial extension of the some outriggers.

(4) The control system sends out an alarm; however, the alarm trips when the capacity is at its limit. The set limit has some margin to prevent crane overturn; however, the alarm comes too late for the operator to try to stop the rotating due to the lowered capacity from partial extension of outriggers.

(5) In other words, the accidents have been caused by "carelessness" of the operator.

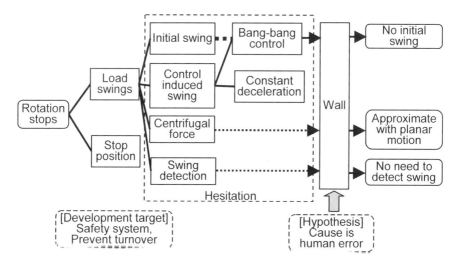

Figure 5.40   Relation of thoughts, hesitation, and hypothesis

Decision: If the above hypotheses were correct, we could model the crane rotary motion and load motion with a plane pendulum motion. The required safety function for the system could be accomplished by automatic rotary motion shutoff, and constant deceleration as in Figure 5.38. The sensor to detect load swing was abandoned because of its low chance of success. We decided to allow some variation in the load stop position and residual swinging motion. We configured the control system with consideration to structural strength, deformation, and response lag in the driving system. Basically, we gave maximum priority to the concept of a safety device that had the full ability to stop the operation.

Results: Starting from the automatic shutoff system technology for rotary motion, we developed other related technologies. In 1991, we commercialized a crane equipped with overload control, which had the automatic shutoff capability against every motion including rotary. Although we limited the automatic shutoff to just stopping, the market gave us high scores to our efforts of putting the automatic control into practice. Simplicity of the system was the key to success. We were the only company that developed the system, though rival companies caught us up with similar technologies as time progressed. The assumption we made for development about the cause of the accidents was finally proven accurate as our system proved effective in preventing overturn accidents.

Summary: We were faced with real doubts during our development. There were concerns about the feasibility of achieving the goal and whether the basic approach was correct, mainly due to limited information about overturn accidents during rotary motion. The development, however, progressed with confidence thanks to establishing correct hypotheses about the accidents.

Sequel: Crane accidents still happen today. From what I have heard, the operators had turned the safety switches off in most cases. We successfully achieved system development to prevent overturn accidents by trusting the operators' skills. The next step is to focus on the operator's mind. After all, it is a human being that operates the machines.

5.1.8 Modified the Lighting Unit of a Wafer Character Recognition Machine – Practice 1

Event: Modified the lighting unit on a wafer character reader to increase its accuracy.

Background: The semiconductor manufacturing process imprints characters on wafer surfaces for identification purposes. The grooves are kept shallow to minimize particles during the processes. The shallow grooves get thin films deposited over them which sometimes are not completely removed in the etching process. Recognizing the characters is thus a challenging task and I wanted to increase the character reader sales by developing a lighting system that read the characters even under such disadvantageous situations.

Functional requirements and constraints: Figure 5.41 shows the analysis of character recognition. Figure 5.42 is the expansion of thoughts diagram for acquiring a clean image. A laser is commonly used for making imprints on wafer surfaces. Figure 5.43 shows the two types of imprints, "soft mark" and "hard mark". The two types have different depths and an automatic character reader has to handle both types. Moreover, the reader has to possess a good recognition probability, even with a resist film or oxide film layered on top of the surface. The original profile of an imprint changes with deposited surfaces. The groove depth and angles are shallower. Colour contrast also differs with the resist film types, thus the reflection coefficient is affected. The reader requires accurate recognition with different wafer states like layered film on top, films etched away, or process liquid on the surface. All these different states should be handled automatically.

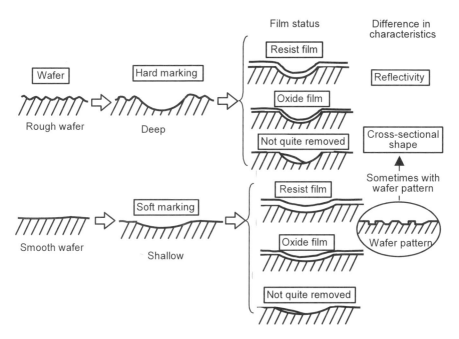

Figure 5.41   Analyzing the object for recognition

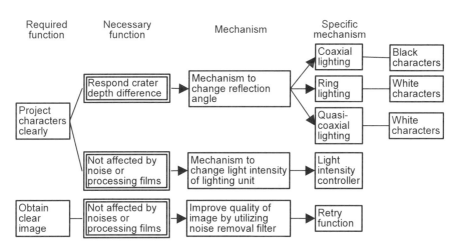

Figure 5.42   Expansion of thoughts diagram for character recognition

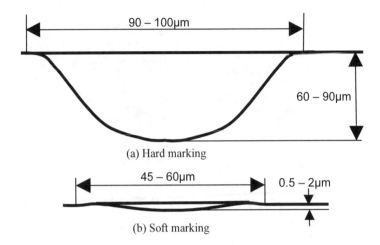

Figure 5.43   Marking types by laser and the cross sectional shape

Evaluation and hesitations: Hard punch imprinting tends to disperse the light due to excessive marking depth. A lighting unit can, however, clear the imprinted images. Types of lighting include coaxial lighting (Figure 5.44) and LED ring lighting which radiates light coaxially or at some angle (10 to 80 degrees). Soft marking, on the other hand, leaves shallow imprints, with a slope angle of approximately 1 to 2 degrees. Lighting from approximately 2 to 4 degrees, therefore, provides sufficient reflectance for the camera to get clear images of the printed characters (Figure 5.46). Placing the light source off the focal point along the focal surface creates angled parallel light at quasi-coaxial lighting, and accordingly, several parallel lights are created if several light sources are placed. Radiating the light at an angle corresponding to the imprint surface angle gives good images when the wafer has layered films or shallow imprints.

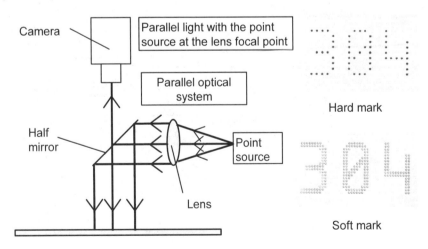

Figure 5.44   Coaxial lighting

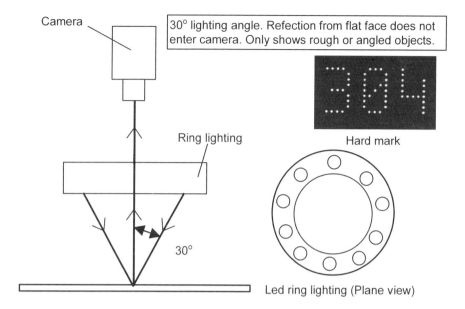

Figure 5.45   LED ring lighting

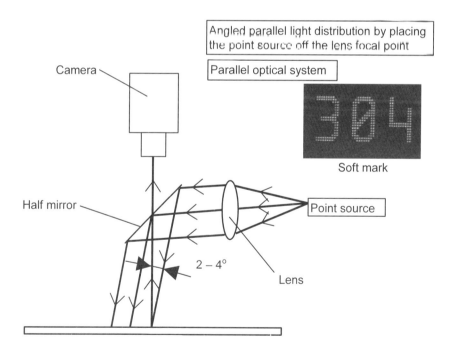

Figure 5.46   Quasi-coaxial lighting

Reflection intensity from imprints varies with the reflection characteristics of the resist film and wafer surface. Blurring occurs when the reflection from the top film is too intense or too weak. A function that controls the light intensity is therefore necessary.

When films or patterns are near the imprint, there is the possibility of the machine totally losing its sight. The lighting unit cannot solve this problem. In these cases, the user has to restrict wafer processing around the characters to keep the area around the imprints clean.

Image filters can remove patterns near imprints. Sometimes the user has to change the parameters of the image filter, depending on pattern type. For the study in this report, it was difficult to change the parameters automatically, and instead, we pre-registered some conditions based on previous samples and tried different conditions. The switching effectively raised the rate of correct recognition.

Decision: We added coaxial lighting, LED ring lighting, and quasi-coaxial lighting to the lighting unit. The light radiation angle for quasi-coaxial lighting was set to 1 to 4 degrees for the larger light angle range. Faces with slopes between 0.5 and 2 degrees gained better visibility for recognizing soft imprints. We made the light intensity from the lighting unit adjustable and recorded the lighting parameters of lighting type, intensity and filter types. We also set up a retrial sequence to raise the chance of automatically recognizing the imprint by seeking different conditions in case of failure with the first set of conditions.

Reasons for decision: For practical use, adjustability upon installation is important. We designed the lighting unit to have only a few movable parts, and thus the installation only calls for setting the unit to squarely face the wafer surface. The recognition algorithm is of course important, however software alone cannot deal with every case. We often have to tune the lighting unit and other peripheral devices for good recognition. Reducing the number of such possible combinations gives the user advantage in constructing an effective system.

Results: We had the system evaluated in Company N's semiconductor lab. Our system proved to have a higher probability of recognition and it was formally accepted by the customer. Even after the order, we registered further reading conditions for the retry sequence to further raise the probability of proper imprint recognition.

Summary: We developed a character reader for wafer imprints with a high probability of character recognition. The system had quasi-coaxial lighting in addition to coaxial and LED ring lightings. The key to our success was the ease in setting up the retry sequence.

Knowledge: In engineering terms, it is important to build the physical model and find the solution to the model. From an operational perspective, the concept of "simple is best" applies.

Related topic: The quasi-coaxial lighting for wafers has the same principle as the hidden-image mirror" the underground Christians used when Christianity was banned in Japan (Figure 5.47). Regular mirror surfaces reflect the light straight and a slight change in the light direction makes the mirror appear dark. The hidden-image or magic mirror brings up an image of Jesus when light hits it at a small angle. Areas of the carved image that point in the right direction shine brightly to produce the image (Figure 5.48).

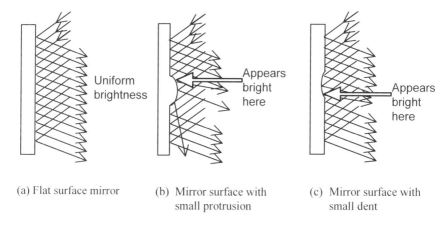

(a) Flat surface mirror

(b) Mirror surface with small protrusion

(c) Mirror surface with small dent

Figure 5.47   Principle of hidden image mirror

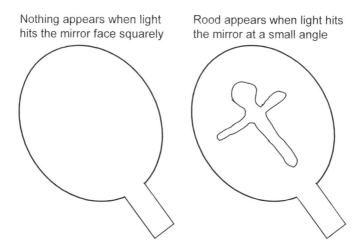

Figure 5.48   How the magic mirror works

5.1.9 Succeeded in Laser Welding by Controlling Its Tip Distance – Practice 2

Event: We developed a system for controlling the tip distance for lap welding with a YAG laser welding system (Figure 5.49).

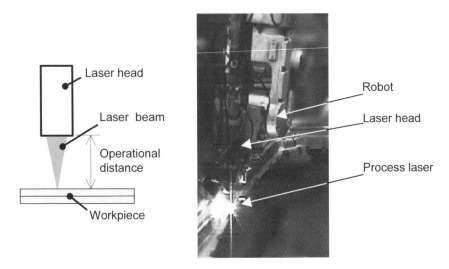

Figure 5.49   YAG laser welding

Background: We had the following requirements for lap welding with YAG laser welding.

    (1) Welding all the way to the back face damages the appearance and we wanted a blind weld like that shown in Figure 5.50(b).

    (2) We wanted to keep high productivity with high speed welding of several meters per minute.

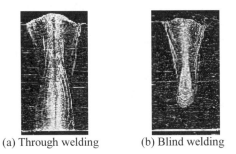

(a) Through welding   (b) Blind welding

Figure 5.50   Cross-section of YAG laser welding

The welding that meets these requirements does not have much margin. Especially, the tip distance has a small tolerance of ±1mm. Controlling the tip distance within this small window is difficult with an open-loop control scheme. Also, when it comes to mass production, a number of factors like positioning the work, the work geometry, and repeatability of the robot's positioning come into play and keeping the tip within a ±1mm range would require feedback control.

A sensor feedback system for this type of work was not available in the market and I decided to build the system myself.

Discussion: The control would be a feedback system with a sensor and actuator. For coping with complex shapes of the work, the actuator would be a six-axis robot.

A number of solutions were available for the sensor to measure the tip distance and I expanded my thoughts as Figure 5.51 shows.

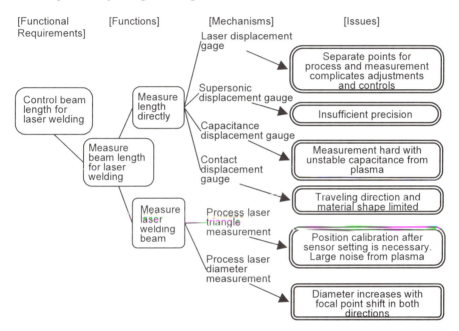

Figure 5.51   Expansion of thoughts diagram for measuring YAG operational distance

First, I thought about using a displacement sensor to directly measure the distance. It would satisfy the function of measuring distance and highly accurate models were available. The point of measurement, however, would not be exactly the point of welding. For handling a large variety of works with different 3D shapes, I thought it would be best to match the point of measurement and the point of welding. If they did not match, I would face the following problems:
 (1) I would need to build a complex software system that coordinated its control with the welding program.
 (2) The position adjustment would be complex, involving the welding point vs. the measurement point, and the laser head vs. the displacement sensor (Figure 5.52).

From the axiomatic design point of view, the system would increase the amount of information, and the information axiom states that such a design is poor.

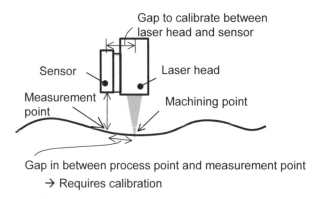

Figure 5.52   Problem with measuring with a displacement sensor

I then thought about using the light from welding, which was emitted directly from the point of welding. There were a number of ways to use the light. The methods were roughly grouped into two categories, as Figure 5.53 shows.

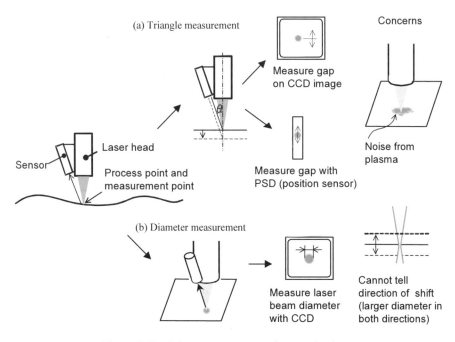

Figure 5.53   Direct measurement of processing beam

(1) Triangle measurement of processing laser: This method installs the sensor (CCD or PSD) on the laser head to watch the processing laser from an angle. It is similar to a laser displacement gauge that uses the processing laser. Although the method requires calibration, the amount of information is minimized since the only calibration is for the gap between the welding point and the sensor. Several concerns, however, came up. The first was

how to quantify the dispersion of the processing laser to other places. The scattered light interferes with the measurement. The second worry was the effect the of thermal deformation of the lenses. Thermal deformation usually affects the focal point distance, causing misalignment.

(2) Processing laser diameter measurement: This method is free of the above concerns. It is, in principle, the best method for measuring the focal point distance itself. No calibration is necessary since a YAG laser allows coaxial observation. This method, however, has the problem that we cannot tell the direction of misalignment from the focal point.

Decision: With the advantage of minimizing the amount of information, we employed the method that directly observes the processing laser in Figure 5.53. Direct observation is equivalent to acquiring information from the point of process. Figure 5.54 shows the decision-making process.

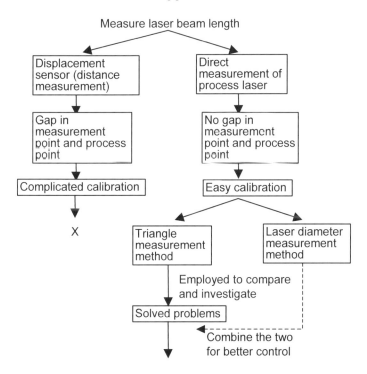

Figure 5.54   Decision process of controlling laser beam length measurement

Results: We designed the triangle measurement method and the diameter measurement method, and built a test system using a CCD camera and PSD (positioning sensor).

Although we had our mind set on the direct diameter measuring method, we achieved satisfactory performance with the triangle measurement method as a feedback control system for compensating the programmed robot movement. This method is easier to control than the direct method and the concerns about plasma light dispersion and thermal deformation of the optical system proved trivial through experiments.

Evaluation: During our evaluation, we found a feedback system was available on the market; however, it was very expensive, and would have required a software change. Our decision to develop a new system turned out to be correct.

Many of the technical elements for triangle measurement were already available and we finished the tests and evaluation in a short time. Adding the direct processing laser diameter measurement to the system might increase the system performance.

Knowledge: A redundant system with too many design parameters, although it may have high accuracy when adjusted right against the number of functional requirements, is not practical because the adjustments are often too complicated. It is necessary to confirm that no additional elements are included in the system when one is designing.

5.1.10 Applied IC Tagging for Managing Metal Mold Parts – Practice 3

Event: We applied radio-frequency identification (RFID) tagging, which stores all the part information, to all the metal parts for managing them. We combined a reader/writer and an information management database to develop the RFID system.

Background: A typical injection-molding die consists of about 200 parts. Every necessary part for a mobile phone takes more than 10 dies and thus over 2,000 die parts are spread out in the factory at the same time. Managing all of them takes enormous efforts. In addition, fabricating each part takes as many as 6 manufacturing processes, and the process control is not an easy task either.

Functional requirements and constraints: Functional requirements were to have each part hold its information, and be able to freely read and write the information. The constraints were how to attach the data media to the diversely shaped die parts, and the media's resistance to harsh environments because they are exposed to oil, water, and heat during the manufacturing processes.

Evaluation and hesitation: Conventionally, part information is printed on paper and attached to the parts for their management. This method, however, is problematic because the paper can easily separate from the part, and moreover, it wastes paper. Figure 5.55 shows how our ideas developed through brainstorming to come up with a new management system.

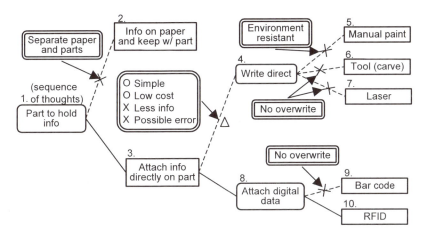

Figure 5.55   Expansion of thoughts diagram

The easiest way to store the information is to write directly onto the part (4). Writing the information sounds simple, but it has several options: apply paint (5), engrave with a tool (6), or mark with laser (7). Method (5) has a reliability problem since the paint is exposed to oil and water. Methods (6) and (7) have a problem that once marked, we cannot change the information. All these methods require human interaction, *e.g.*, transferring the information from the paper to the database. Attaching digital data onto parts (8), *e g*, bar-coding, can avoid such problems. Labeling parts with barcodes avoids human error, which is quite common with manual information processing, and accesses the database in a fraction of a second. The downside of using a barcode is the same with methods (6) and (7), *i.e.*, once marked, changing the information is not trivial and since the barcode optically provides information, stains can cause malfunction. Another problem is, that methods (4) and (9) can store only limited amount of information, so their usage is limited to maybe managing part IDs. We then came up with a new idea of storing information on RFID tags (10).

Many of us have already encountered RFID technology with automated gate systems for chairlifts at ski resorts. The system allows skiers holding a square plastic chip of about 20 x 20 x 4mm to pass the gate. RFID technology can read and write up to several kB of information on an RFID tag, a non-electrically powered chip, without contact. Some RFID tags are as small as 2mm in diameter and 10mm long. We then set out to find ways to utilize RFID tags on die parts. RFID tags would satisfy the functional requirement; however, we were concerned with cost and reliability. A bigger worry was the possibility of disturbance in the magnetic field with a metal part near the RFID tag and failure to properly transfer the information.

Decision: We decided to use RFID tags for marking information on die parts.

We found a special type of RFID tag that is not affected by the presence of metal nearby. It has been in use for managing manhole covers. This special type is not affected in its information read/write even when it is attached to metal parts.

We also ran some tests to find the its durability and the effect of electromagnetic noise from the factory machines on it and cleared both concerns. RFID tags cost a few dollars each, but we can reuse them over and over. Except for the initial investment, therefore, cost was not a big concern.

Process and results: We attached RFID tags to all the metal parts, and placed a number of reader/writer (antenna to transfer data on the tags) units in the material warehouse, part warehouse, and all the machine tools. All information read by the reader transferred instantly to the database. Figure 5.56 shows the computer system that manage all the processes for all the parts. The system showed what part was at which process and this information was readily available at any information terminal. We finally eliminated all human errors, of part mishandling, process skip, and other part management mistakes.

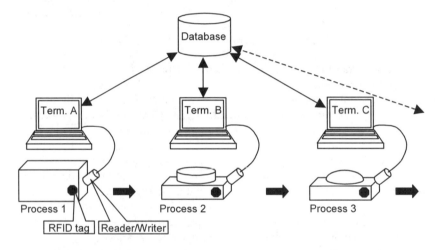

Figure 5.56   Structure of the system

Afterthoughts: Where to store the part data is a critical issue for part management. Typical methods attach only the ID to parts and store the data in a database on a computer hard disk. RFID can store, in addition to the ID, product name, customer information, delivery date, how far it has gone through the entire production process, and so on. Without having to access the database, operators with handy-readers can check the status and what the next process is for a part just lying in the factory. Such convenience could not have been accomplished without using RFID tags.

Knowledge: Information by itself is useless and it has to be read, used, and written when needed, in the form required. The method of reading and writing has to be simple and error-free. Information on paper is read by the human eye, thus it always has the possibility of being subject to human error. When all information is passed back and forth within the machine or a system of them, paperless fabrication can be realized.

5.1.11 Modified a Sandblast Machine for PDP Class Substrate Machining – Practice 4

Event: We resolved a technical project given to us by a plasma display panel (PDP) manufacturer, A, and won their order for capital investment. We accomplished the goal by merging distinct technologies from our related companies to propose a challenging and cost-effective solution. Through the course of winning the order, we overcame a technical difficulty with a well-planned verification test.

Background: It was the time when PDP was under the spotlight as the wall-type television to replace the CRT monitor. Every manufacturer was seeking a method of mass production and starting to build its main factory for PDP production; and the mass media were reporting big stories about the heat. Company B that I work for had a small possibility of building its own PDP factory and its proprietary technologies about microfabrication were about to go down the drain if it could not sell them to any PDP manufacturer.

Functional requirements and constraints: PDP glass substrate manufacturing requires the forming of precise partition walls over a 50 x 50 inch area on the substrate within 3 minutes. Our competitor C was well established in the field and I had to come up with a distinct advantage in the cost of building the process machine.

At the time, my company, B, owned distinguished technologies in quantitative feeding and discharge nozzles for powder in microfabrication. Company B also had an affiliation with Company D which possessed high technology in sandblasting and recycling powder. At the time building upon the joint development by D and B had no chance of beating the cost of competitor C's work, and to make the story even worse, the management of B's production group had cut the number of members from both D and B in this project from five down to one, just myself. My assignment was to work with Company D to devise, propose, and complete a cost-competitive new technology.

We at that time, however, had the competitive edge in spray nozzles, which govern the quality of partition formation by sandblasting. Figure 5.57 shows the rectangular spray nozzle that we had developed. Round nozzles were more common in sandblasting at the time. For a machine, however, that requires cutting 200μm wide and 150μm deep grooves in a direction perpendicular to the PDP screen (and their number should match the number of horizontal pixels), rectangular nozzles had the advantage. The reason was that compressed-air flow forms a uniform flow along the nozzle array when they are rectangular. Blowing fine particles of diameter 20μm in diameter forms the partitions on the glass substrate which is covered with stripes of photosensitive resist. The machining tolerance at the bottom of the partitions wwa within a few percent.

Mind process: To win the order, we needed a fundamental cost reduction plan. I then concentrated on the powder mixer with a relatively large cost factor. Figure 5.58 shows the cross-section of the existing powder mixer. The compressed air, with its volumetric flow rate controlled, flows through the exhaust. The powder is discharged by the screw perpendicularly into the air flow and is carried to the nozzles. The container is pressurized to a degree subject to the compressed flow volume through the pressurization path. This powder mixer governs the performance of sandblasting. Our well-planned, cost-effective proposal was to install multiple screws and nozzles in a single pressurized container. Figure 5.59 shows the three-dimensional view of the mixer I proposed.

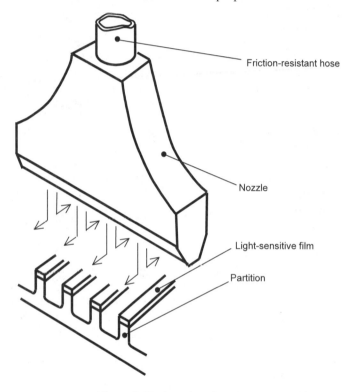

Figure 5.57   Nozzle and partitions

There was no precedent for the new machine; however I thought it must be appealing to Company A because it had multiples heads (=nozzles) on a single body (=pressurized container) to significantly cut down the cost.

Without time for basic experiments, the sales activities took off. Matching the volumetric flow rate for the multiple air paths and the pressure loss would keep the same machine performance, but the world of powder sometimes plays tricks against reasonable guesses, and I thought I had to prove the principle of my plan for cutting the cost.

Realization: The concern I had about installing multiple screws and nozzles on a single pressurized container was fluctuation in the powder spray due to interference of screws and nozzles controlled individually. The concern would be gone if I proved that the amount of sprayed powder depended only on the rotational speed of the screw and was irrelevant to the compressed-air flow rate. I then took the jointly developed machine apart and put the pressurized container on an electric balance to measure its rate of change in weight when the powder was being discharged. This may have been an outdated way of making a measurement, but it would tell the truth for sure. Fortunately, Company D had purchased a digital scale the previous year, and it was still sitting in our warehouse. I carefully designed the test setup to take away the stiffness from the hoses at junctions and even requested help from those experienced in using the electrical balance.

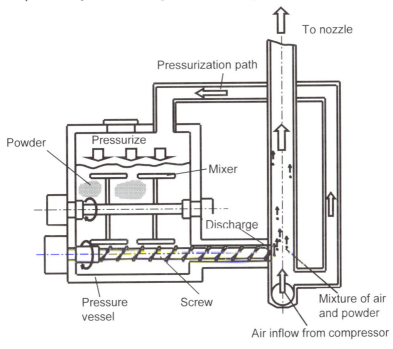

Figure 5.58   Cross section of conventional powder mixer

Despite the concerns, the test was a success. The powder spray remained constant whether the compressed-air flow volume was at maximum or zero. The weight of the entire container changed linearly, showing its independence from changes in the compressed-air flow volume or pressure. Later when we delivered the machine, we faced a minor problem; however, thanks to this test run, I quickly fixed the problem in a short time with confidence.

Knowledge: If there was no time to verify or demonstrate your idea because of the delivery schedule, search for ways to prove the basic concept even though it may seem trivial. It is a place you can always go back to if there was some misunderstanding and you may even get results better than your expectations.

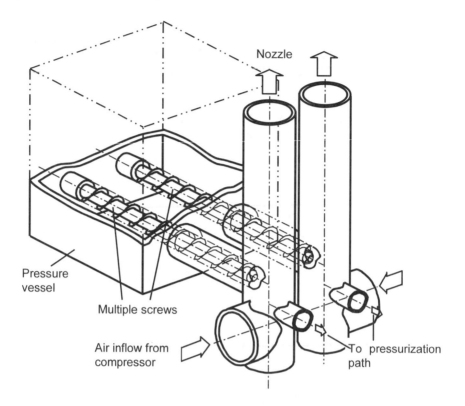

Figure 5.59   Powder mixing machine I proposed

5.1.12 Installed a New Cable on a Multi-Joint Robot – Modification 1

Event: In order to install a plasma cutting torch to a multi-joint robot, we discarded the existing cable and installed a new torch cable designed for use on the multi-joint robot.

A plasma cutter cuts metal by discharging plasma gas from the tip of the torch. The process applies high-frequency high voltage to air and oxygen going through the torch. The continuous direct current retains the plasma state.

Background: Improved precision of multi-joint robots, and its cooperative control with the plasma cutter, allowed the precise cutting of oddly shaped three-dimensional objects. The industry had started using more multi-joint robots with plasma cutters. The plasma cutter may not be as precise as a laser cutter, however its easy operation, less peripheral equipment compared with gas cutters, low cost, and looser safety regulations have contributed to the increase of applying plasma cutters on multi-joint robots.

Course: The torch of a plasma cutting machine is installed on a flat-face table. With the torch mounted on a multi-joint robot (Figure 5.60), the three-dimensional movement of the robot arm puts stress on the torch cables by frequent bending, and friction at the elbow (Figure 5.61). We decided to redesign the existing torch cable.

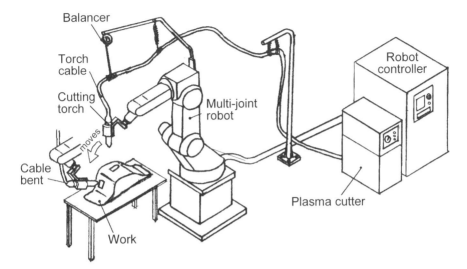

Figure 5.60 Structure of the cutting robot and where the cable bends

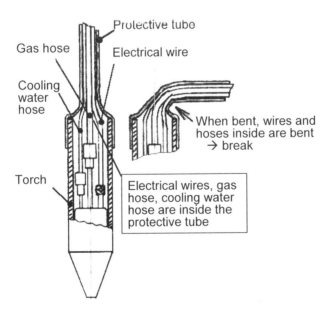

Figure 5.61 Conventional torch and cable connection with sharp bend

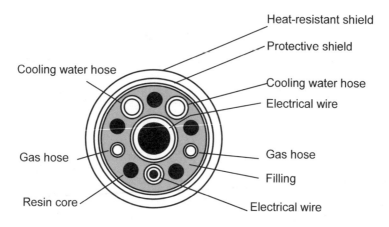

Figure 5.62   Cross section of integrated cable customized for the new design

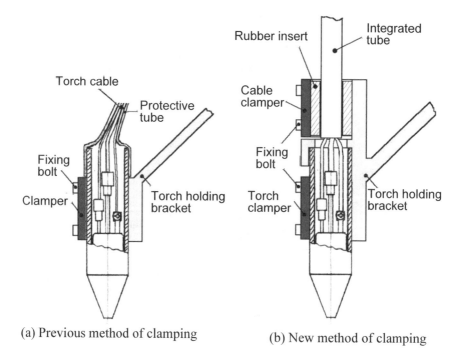

(a) Previous method of clamping   (b) New method of clamping

Figure 5.63   Old and new torch cable clamping

Countermeasure: We installed a cable specially designed for use with the multi-joint robot. The cable itself has a certain amount of rigidity (Figure 5.62). We also installed a cable clamp on the torch bracket to rigidly hold the cable end. This clamp prevents the cable from bending at its end (Figure 5.63). We ordered a long cable (a year's supply) from the cable manufacturer to minimize the negative impact of cost increase. Our modification improved the overall market value.

Mind process: We at first wondered if we could modify the original clamping with the configuration at the time. Ideas we had included adjusting the suspending of the cable and wrapping it with reinforcing tubes. We quickly found that the cable bending motion could not be avoided. Also, reinforcing the cable simply moved the bending to another location.

Limiting the robot work volume or speed together with the above changes would prevent cable damage; however, such restrictions would lower the robot's market value. We then evaluated a new cable with flexible bending and a certain amount of rigidity (Figure 5.64).

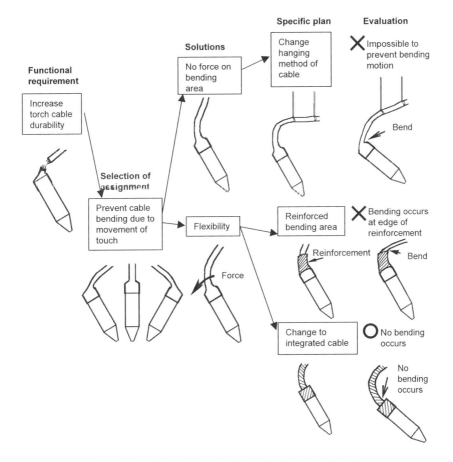

Figure 5.64   Expansion of thoughts for the torch cable

Although it was just a cable, a new design would involve some development and we were concerned with unexpected side effects like loss of a good plasma or a drop in reliability. Introducing a new cable design would increase the price, and require revising the instruction, inspection, and service manuals. Also, parts for production and service would increase as well.

In fact, early wear-out due to bending is not a big problem compared with customer complaints from the market, and the user manual states that the cables are consumable parts. It came to our mind to write a manual for part replacement and keep a good supply. Also, consumable parts have large profit margins and it was tempting to keep the design at the time (Figure 6.65).

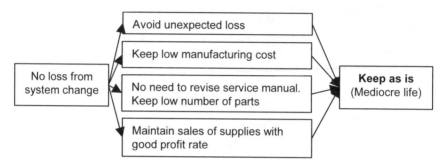

Figure 5.65   Hesitations about modifying the torch cable

In the long run, however, we came to our senses and recognized that products that give a hard time to the customer would eventually cost us the loss of the client, and we decided to develop a new cable design for the multi-joint robot (Figure 5.66).

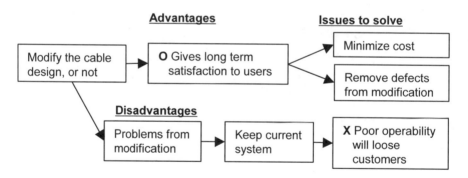

Figure 5.66   Mind process to reach decision to modify the cable design

Once we made the decision, the rest was to carry out a cost-effective project to develop the new cable without introducing troubles from the change. We determined the structure and dimensions of the cable by checking the plasma performance, and balancing the design in terms of rigidity and flexibility. We kept the unit cost low by ordering a length of cable that would last for a year (several hundred meters) and would minimize the effect on cost (Figure 5.67). We checked how reliability was affected with repeated bending tests and durability tests, and at the same time looked for any unexpected side effects. We then found that the heavier cable made the time for the torch to stabilize after moving a second longer (hence a drop in the cutting process productivity). We judged that the effect of this longer wait was negligible in terms of the entire work process and that the shortcomings for the user were small.

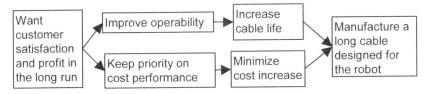

Figure 5.67 Requirements and action for selecting integrated cable

Figure 5.68 shows the various thoughts that went through our minds during the development.

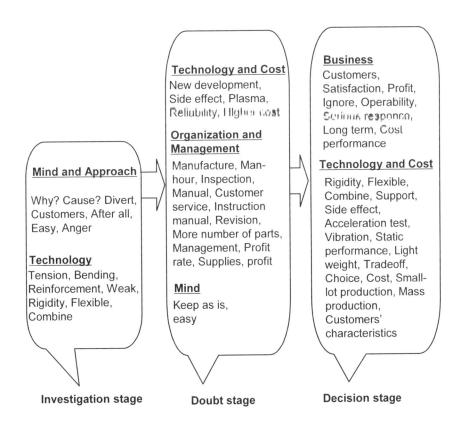

Figure 5.68 Mind process from start to improvement

Knowledge:

- Adding a patch fails in the long run.
- Diverting to an already available solution costs more in the end.
- Being too nervous about cost increase leads to customer dissatisfaction. Look at the cost cut vs. the negative impact.
- Cost of parts for small-batch products can approach mass production part cost by producing a stock of the minimum necessary quantity.
- A short-term profit can lead to a long-term loss.

Sequel: The special cable for robots was well accepted by the market except for one customer who rejected it because of the longer stabilizing time. This customer was using the conventional type of cable.

Conventional plasma cutting machines that cut planar steel plates on a flat table were later required to make angled cuts and started to act just like multi-joint robots. We started to put the cables we developed for multi-joint robots on regular plasma cutting machines. The part is now common to both types.

5.1.13 Modified a Material Cooling System but Could Not Cut the Cost – Modification 2

Event: Attempted to cut down the cost of cooling material by shortening the cycle time but the cost did not drop.

Course: We restructured the material processing line. Running production at the target cycle time would result in doubling the energy cost of cooling the material. The cooling equipment had the constraints to fit in a preassigned space and cool the parts within the target cycle time. To increase the heat extraction rate, we had to lower the temperature of the cooling fluid. We used a chiller to cool the cooling air. Cost calculation showed a large electricity usage and the cost overshot the target product cost. We decided to have a longer natural cooling time to solve the running cost issue (Figure 5.69).

Cause: We jumped into modifying the equipment without listing all the functional requirements.

Given a project, we tend to all look at just the project. The designer was only concerned with increasing the efficiency of the equipment to cool the material in a short time. We had a large number of development items on the table and not enough time for thorough discussion. All those in charge worked on shortening the line time; however, each had his own idea and it was not clear who was in charge of keeping an eye on the overall process.

Action: We took on the restructuring without looking at the whole picture. Midway through the project, we figured that the target performance could not be met; however, it was too late to change the entire project. We continued with the project and added a chiller. We made no cost savings at all.

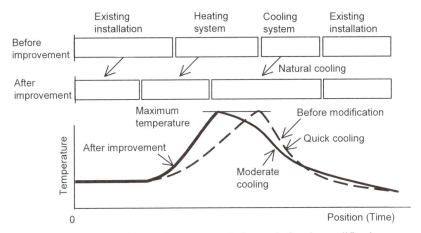

Figure 5.69   Position and temperature before and after the modification

Knowledge:

- Think from scratch when you design equipment. Thinking about extending the existing equipment can lead to failure.
- Think what determines the cost and think flexibly.
- Clarify who is the leader to watch the overall progress.
- Not seeing the wood for the trees is the worst failing.

5.1.14 Successfully Anchored a Floating DNA Fiber in Liquid – Research 1

Event: Developed a method to anchor a microscopic DNA fiber floating in liquid onto a glass substrate using static electricity.

Background: Conventional biogenetics statistically handle DNA in liquid, for example with electrophoresis. In my master's degree studies I had already been developing a completely new way of handling DNA as Figure 5.70 shows. This systems cuts out specific parts of DNA with a microscopic mechanical probe and separates them on another clean glass substrate.

DNA is a macromolecule fiber of diameter approximately 2nm and generally floats in a solvent. Observing or handling DNA in such a state, I needed to develop a way to anchor the floating DNA to a glass substrate.

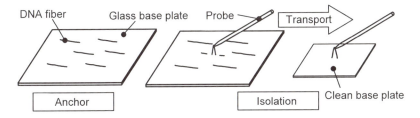

Figure 5.70   Handling system for a single DNA molecule

Functional requirement and constraint: The function needed was to anchor a DNA on a glass substrate. As Figure 5.71 shows, the DNA is observed by an inverted fluorescence microscope with its objective lens under the substrate. Observation is done through the substrate and from underneath it. Thus there was the constraint that the glass substrate to hold the DNA had to be transparent.

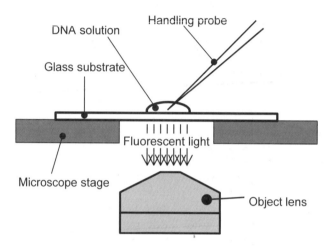

Figure 5.71   Observing DNA with an inverted fluorescence microscope

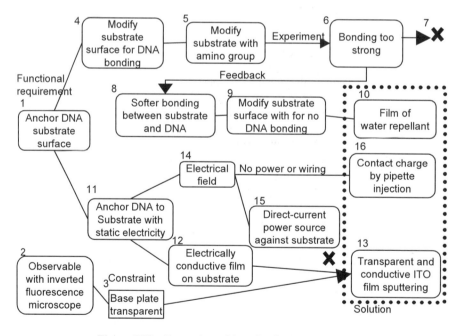

Figure 5.72   Expansion of thoughts for DNA anchoring

Mind process and hesitation: The expansion of thoughts diagram in Figure 5.72 shows my mind process to reach a decision together with my hesitations.

(1) I first thought of coating the substrate with a substance to chemically bond the DNA. I immediately ran a test without the time-consuming theoretical calculation. I first prepared an amino-group-coated glass substrate because the amino group bonds well with DNA. I dropped some DNA solution on the surface to make an observation. I then confirmed that the DNA was successfully anchored.   (1→4→5)

(2) I attempted to mechanically handle the anchored DNA with the tip of a microscopic probe. The DNA, however, was bonded firmly to the substrate and I could not pick it up. It was then that I discovered the new functional requirement to control the strength of the DNA bonding to the substrate.   (6→7)

(3) I then thought about coating the substrate with a substance which is hard to bond a DNA fiber to. I applied a membrane coating with water repellant of high density and high viscosity.   (8→9→10)

(4) Water repellant made it impossible for DNA to bond to the substrate with chemical bonding force, and I therefore moved my thoughts to static electricity bonding that works from a distance. The DNA has a single phosphoric group on a single base, and it ionizes in the solution to function as an anion. I therefore thought about anchoring down the DNA by adding electrostatic force in the direction towards the substrate by producing an electric field.   (1→11→14→15)

(5) I then needed to install an electrode on the substrate surface to anchor the DNA with static electricity. The last thought was to form a conductive membrane on the substrate surface. Metal is the most common conductive material, but it is not transparent and blocks the view for the inverted fluorescence microscope. When I was searching for a conductive and transparent substance, one of the senior associates from the master's program told me about ITO (indium tin oxide). ITO is well known as a substance for as the transparent electrode. Immediately, we sputtered an ITO membrane on the substrate. I then hypothesized that setting a planar electrode facing the substrate and applying voltage between the two electrodes could anchor down the DNA to the substrate surface.
                                                                                                           (11→12→13)
                                                                                                           (2→3→13)

Realization and results: I created an ITO conductive membrane on the substrate surface, and a water-repellant membrane on top of the first layer. Then I dropped a DNA solution onto the second layer, and placed an electrode plate on top of it. When the two electrodes were energized, they anchored down the DNA to the substrate.

This accomplished the purpose of anchoring down the DNA. I discovered, however, as the experiment progressed, that energizing the electrodes was not really necessary for the anchoring. The process could be achieved by placing conductive material over the lower substrate surface. Figure 5.73 shows my reasoning that a negative electricity charge was formed by rubbing the DNA solution during the DNA ejection from a polypropylene pipette. In conclusion, I anchored down DNA to the glass substrate, by simply dropping the DNA solution onto the substrate coated with a membrane of ITO and water repellant. Figure 5.74 is an image of anchored DNA captured by the inverted fluorescence microscope. The discovery reduced the cost and labor of the project, and led the development to a total success. The actual handling experiment confirmed that the DNA bonding strength to the glass substrate was right on target.

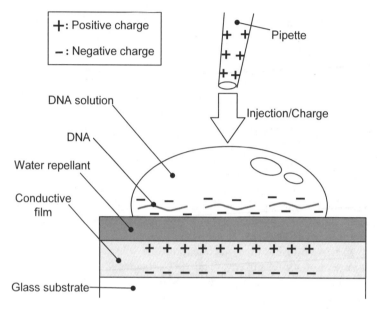

Figure 5.73   Principle of DNA anchoring

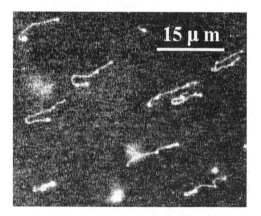

Figure 5.74   Inverted fluorescence microscope image of DNA anchored to substrate

Discussion and evaluation: I found the DNA bonding strength to the glass substrate too strong with the first experiment. This was and example of running the experiment first before calculation could provide useful information. Feedback from this test directed the decision-making. Figure 5.75 shows the spiral diagram of the decision process.

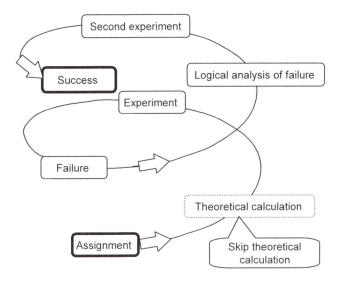

Figure 5.75   Spiral diagram showing the decision-making process

In addition, the mind process to reach ITO started from Anchoring DNA → Electrostatic force → Electrode → Conductivity, and by the added constraint of conductive and transparent, → Transparent electrode (ITO). Clarifying the functional requirements and constraints with the expansion of thoughts diagram led to the good idea.

Knowledge:
(1) Carrying out the experiment first before calculating can sometimes lead to simplifying the decision process when you have several ideas.
(2) Clarifying the functional requirements and constraints with the expansion of thoughts diagram leads to good ideas.

Related thoughts: The reason for the DNA to make a strong bonding to the glass substrate was probably intermolecular force. This reminds me of the difficulty of getting a stain off a plastic container if the stain got on there when it was dry. Recently, you may have noticed that many gentlemen's urinals flush a small amount of water when someone stands in front of them. They place films of water on the urinal beforehand so that molecules from the urine do not get stuck to the surface.

5.1.15 Made DNA with Fluorescent Molecule Visible in Liquid – Research 2

Event: We mixed DNA with fluorescent molecules, which bonded to specific parts of DNA, in liquid to observe the intended parts of DNA. After mixing, the entire liquid glowed so we came up with a way of separating unwanted fluorescent molecules from DNA. We used DNA extracted from a lambda phage virus (2nm in diameter, 15μm in length).

Background: DNA fragmentation can lead to the decoding of its base sequence. The conventional method is to use a restricted enzyme to chemically fragment DNA. This method, however, chops the DNA into pieces at multiple locations and the original position information is lost. Reconstructing the original form takes an enormous amount of time and manpower.

Today, a method that mechanically cuts a piece of DNA at one place with a sharp object has been developed. The method has the advantage that it does not require the reconstruction of the entire piece from fragments and allows faster DNA analysis. The method, however, cannot tell the exact location or orientation of the cut. We therefore, as an alternative, decided to mark targeted positions on DNA.

The technology exists for bonding fluorescent molecules to a target location on DNA. First the DNA is dried and fixed to a glass substrate. Then the entire substrate is submerged under a solution that contains the fluorescent molecules. At this point, the fluorescent molecules automatically bond themselves to the target position on the DNA. This behavior is caused by the fluorescent molecule structure pairing up with the particular portions of DNA. The glass substrate is then washed with water to to remove the excess fluorescent molecules (Figure 5.76). The method, however, generates a strong bonding between the DNA and the glass substrate, making further operations like cutting impossible.

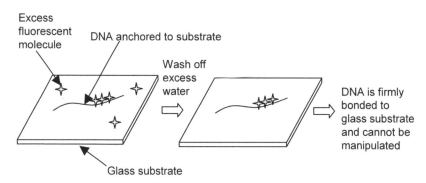

Figure 5.76   Existing method of marking DNA

Given the above, I was trying to develop a new DNA visualizing technique for randomly marking the target positions while maintaining further operability.

If we mixed the DNA and the fluorescent molecules in liquid, the fluorescent molecules would bond to specific portions of the DNA without fixing the DNA to the substrate. This approach could not, however, wash away unwanted fluorescent molecules with the DNA not fixed to the substrate (Figure 5.77). Thus, I needed to develop a way to separate the DNA and unwanted fluorescent molecules in the liquid.

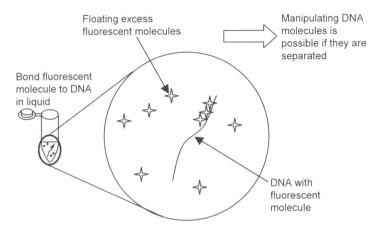

Floating excess fluorescent molecules

Manipulating DNA molecules is possible if they are separated

Bond fluorescent molecule to DNA in liquid

DNA with fluorescent molecule

Figure 5.77   DNA floating in liquid

Functional requirement and constraint: The functional requirement was a way of removing unwanted fluorescent molecules from liquid where DNA and fluorescent molecules attached to specific locations on the DNA were also present. The study was aimed at analyzing the base array, so the collection rate and reliability of the purity were constraints. Also, to achieve faster analysis, the process should be simple and quick.

Mind process: Figure 5.78 shows the expansion of thoughts diagram for reaching the decision. I first focused on the molecular difference between DNA and fluorescent molecules, and noted the difference in the density and molecular mass (=size).

(1) First, I attempted a method using density to separate the DNA from fluorescent molecules. A DNA molecule is much larger than a fluorescent molecule, and their mixture when subject to centrifugation would precipitate DNA, fluorescent molecules, and solution in this order (Figure 5.79). DNA and fluorescent molecules, however, are achromatic and I could not distinguish the boundary between them. I attempted to pick up the precipitated DNA from the bottom by pipette, but the method clearly lacked accuracy.

(2) I then tried electrophoresis. The method buries the mixture in polymer gel and applies an electrical field to it. Both DNA and fluorescent molecules have negative charges and thus are attracted to the positive side. The gel has a network of fibers and thus the smaller fluorescent molecules will travel faster than the DNA with its large molecules. This method, however, was time-consuming because each electrophoresis process took about an hour and I had to repeat it several times. This method was not the solution either.

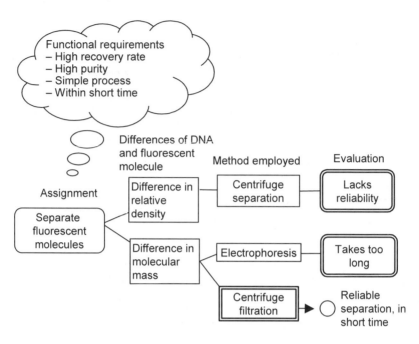

Figure 5.78   Expansion of thoughts for fluorescent molecule separation

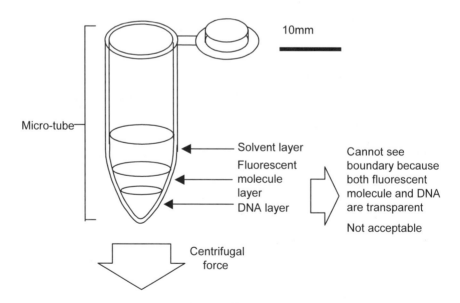

Figure 5.79   Centrifuge separation method

(3) The next method took advantage of the difference in molecular size between the DNA and fluorescent molecules. Figure 5.80 shows the micro-tube we used. The tube has a filter membrane that only passes the small fluorescent molecules whereas it blocks the large DNA molecules. The amount of sample I had was approximately 0.1cc and gravitational force was not strong enough to break the surface tension between the liquid and the inner surface of the micro-tube. I then forced the filtration with centrifugation. Each centrifugation takes only about 3 minutes and I only had to repeat it several times. This process was simple.

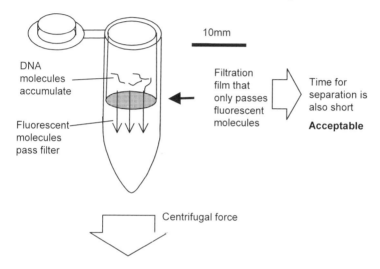

Figure 5.80   Centrifuge filtration method

Results: We prepared a micro-tube with a filtration membrane inside. We placed the micro-tube in the centrifugal separator after pouring the liquid with DNA and fluorescent molecules into the micro-tube and activated the centrifugation. Once the process was over, we recovered the liquid on the membrane and observed it with a fluorescent microscope. Figure 5.81 shows the fiber-shaped fluorescent light we saw through the microscope. If we scoped the mixture, the entire view was bright with fluorescent light. The fiber-shaped bright image in Figure 5.81 proves that we succeeded in removing the unwanted fluorescent molecules from the mixture.

Knowledge:
- Clarifying the functional requirement and carefully observing the phenomenon sometimes leads to the solution.
- When you have a number of possible solutions, actually trying each method will tell you if it is right or wrong.

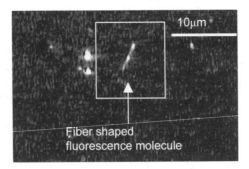

Figure 5.81   Observing recovered liquid with fluorescence microscope

5.1.16 Built a Microscopic Assembly Tool – Research 3

Event: We invented a tool that can grasp and release a microscopic object with static charge. The tool allows microscopic assembly under an electron microscope.

Background: We, with a university laboratory, were conducting a research to assemble microscopic objects with a manipulator under an electron microscope. Similar researches were underway at some other places as well. Through the electron microscope we can observe objects with resolution of under 1μm. The manipulator we developed has a total of 20 degrees of freedom in traversal and rotational movements. The operations, however, were limited to lining up the fine parts since the tool that actually carries out the assembly was missing.

Functional requirement and constraint: The functional requirement for the assembly tool was the grasping and releasing of the physical objects made of conductive material in the size range between 1μm and 100μm. Once this requirement was met, the manipulator would be able to build assemblies, because assembly is a combination of relative positioning among objects. The assembly process also requires push and pull functions, but they could be achieved by combining "grasp" and "move" functions with the assembly tool. Constraints were that use of liquid or gaseous matter was prohibited for the assembly tool because the mass values of physical objects to be handled were extremely small, and the use of an electron microscope required a vacuum.

Mind process: We started by narrowing down the alternative mechanisms based on the constraints. The method we used to make designs was to settle with the first mechanism Figure 5.82 and to immediately prototype it. This time, however, we selected the mechanism to use electrostatic force, then instead of immediately going to prototype building, we ran some virtual experiments (or an actual experiment) against new constraints and the fabrication methods shown below the double dotted line in Figure 5.82. We repeatedly went back and forth between the tentative design and virtual experiments to reach the final design solution.

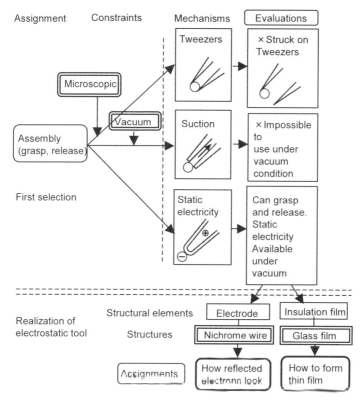

Figure 5.82   Expansion of thoughts for selecting the electrostatic tool

The alternatives we came up with were microscopic tweezers, and a suction tool that grasped the object at the end of its narrow glass tube. The suction tool was infeasible with the vacuum constraint, and the micro-tweezers had the shortcomings of difficulty in grasping extremely small objects without squashing them, and also of difficulty in releasing objects that tended to stick on them. The sticking was due to the strong electrostatic force on the object charged by the electron beam radiation, and surface tension from the small amount of moisture and oil in the vacuum. We then thought we could control these forces to accomplish the grasping and releasing. In the end, we developed an assembly tool that used static charge (Figure 5.83).

Figure 5.84 shows the expansion of thoughts diagram for designing the assembly tool. We had to cover the fine electrodes with thin and uniform insulation film. Such a film, however, is not easy to create, and we made several attempts using insulation paint or aluminum films, which all failed. We also tried to fill the glass tube with conductive paste but it did not reach the narrow end. We were also concerned that the electron microscope image receive noise.

Decision: Although we were concerned about noise on the image of the electron microscope, we set our minds on the electrostatic tool to grasp and release objects. We started to plan how to produce it.

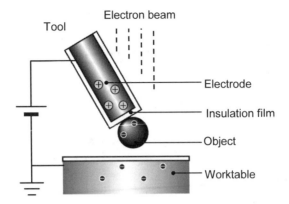

Figure 5.83   Principle of electrostatic tool

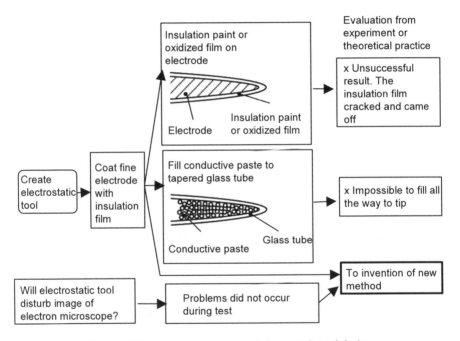

Figure 5.84   Evaluation process of electrostatic tool design

Design: Figure 5.85 shows the new process we came up with to create a metal bar with a uniform glass insulating film. The process started with a nichrome wire passed through a glass tube and we locally heated it, tore it where it was heated, polished the tip, and then reheated the tip. We took a hint from the optical fiber pulling process from raw material. Figure 5.86 shows the actual manufacturing process. During the tearing process, the nichrome wire inside the hollow glass tube broke before the glass tube did, and the glass tube extended a little further before it broke as well. The sanding and reheating melted the extended glass which then covered the nichrome wire tip.

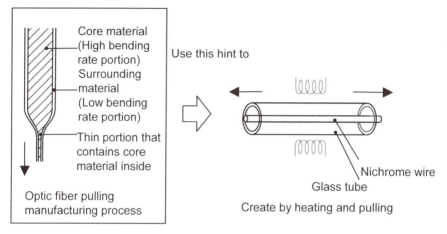

Figure 5.85   New method for producing electrostatic tool

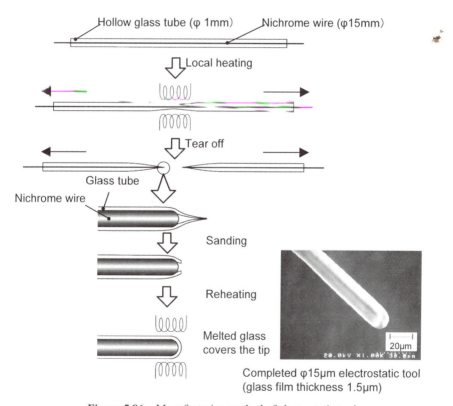

Figure 5.86   Manufacturing method of electrostatic tool

We reduced the effect of the electric charge in the tool by lowering the voltage of the electron microscope detector, which collected the secondary electrons (the electrons discharged from the sample when exposed to electron beam) and increasing the detected component of the reflected electron. The tool electric charge, however, affected the electron discharge and positively charged tools appeared dark, whereas negatively charged ones appeared white.

Results: We developed a glass-coated metal bar (diameter 15–100μm) to grasp and release micro-objects (sphere, cylinder, or plate) of size 1–100μm. Under observation with an electron microscope, the object was charged negatively with the electron beam, thus, we could grasp it by giving the tool a positive charge, and reversing the electric charge released the object from the tool to the workbench.

Discussion: The object that attached to the tool stabilized at the position and orientation where the static charge was maximized. If we formed a spot on the tool where the glass insulation film was thin, the object attached itself there because electrostatic force was inversely proportional to the distance between objects (Figure 5.87). The electrostatic force was also proportional to the contact area, and a flat area on the tool made it easier for the object to attach to. Given this discussion, we actually made a tool with a flat tip and insulation on the area thinner than the sidewalls and observed that objects attached to the tip easily.

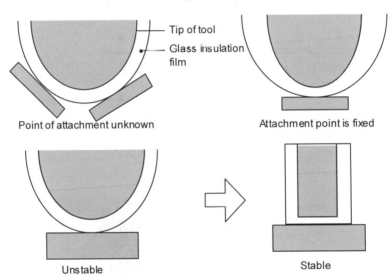

Figure 5.87 Objects attaching to tools and the stability of attachment

Knowledge: In the microscopic world, gravity or inertia, which are both proportional to the volume, are negligible and forces that are proportional to the area, like electrostatic force, are more effective in manipulating objects. Static electricity is usually a nuisance that attracts and fixes dust, but we can make good use of it if we use it right. This is a good example of "exchange/reverse" of the brain operation. The fabrication of the electrostatic tool, when it was needed to attach an insulating film to the outside, consisted of first placing metal inside an insulating tube and then pulling the whole assembly. This method is another example of the mind operation of "inverse idea" and "prepare first."

Related thoughts: When we viewed the tool through the electron microscope, the black-and-white brightness shifted depending on the positive and negative electrical charge.   Like thermography that displays temperature by colors, it helps to clearly visualize the electrical potential, which the naked eye cannot see, through the electron microscope. We took the idea of pulling the glass tube from the production process of optic fibers, an idea which shares its principle with the making of Kintaro candy (Figure 5.88), a traditional Japanese candy with the legendary character Kitaro's face (you can cut the candy anywhere and the same face shows on the section). This principle was to extend a thick object by pulling it to produce a desired section.

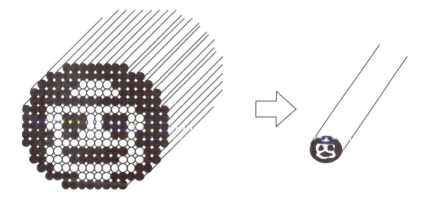

Figure 5.88   How Kintaro candy is made

5.1.17 Accomplished Wide-Range High-Precision Positioning – Research 4

Event: For a read/write testing machine for optical disks, we developed a dual-stage actuator with a wide tracking range. The tool accommodated an optical disk under development with a large eccentricity.

Background: We were engaged in the development of the next-generation optical disk which would double the capacity of conventional disks. I was in charge of evaluating the read/write performance. The media development group kept sending in new disks with different parameters and I had to run them through the evaluation test effectively.

Products have small amounts of eccentricity, but disks from the media development group often had "centers" that were off by amounts several times those of the products. Such large eccentricity caused the tracking device to lose track and my tests were losing efficiency.

Functional requirement: Figure 5.89 shows how the optical disk test machine was structured. An optical disk has a spiral groove and the laser beam forms record marks along tracks between the grooves. If the disk is eccentric, its rotation causes large waves on the track. A galvanometer mirror tilts to control the laser beam to follow the center of the tracks (tracking).

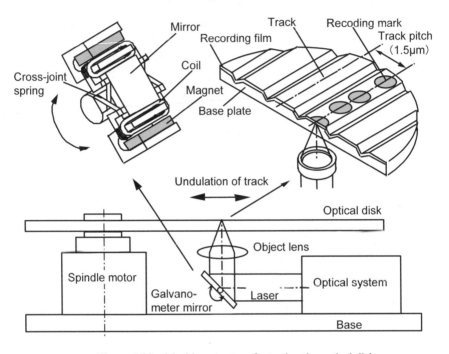

Figure 5.89   Machine structure for testing the optical disk

The range a galvanometer mirror can handle is usually about 15μm pp (peak to peak). Some of the sample disks that I received had eccentricities as large as 300μm pp. Figure 5.90 shows the eccentricity vibration characteristics (spectrum of wave motion) vs. those of the ganvanometer mirror tracking. The tracking control had to control the waving to under 0.1μm. The area shaded with angled lines is the controllable region, and the figure shows that a 50Hz 300μm pp spectrum is out of the control range.

Countermeasure: First I watched the track's eccentricity with a microscope as I mounted each disk on the spindle, so it would turn without eccentricity. Since I had to make this adjustment every time I collected data, my test efficiency was very poor. Then I tried adjusting the galvanometer mirror to widen its range to use different types of actuator with a wider range, but they could not handle the large range of 300μm pp.

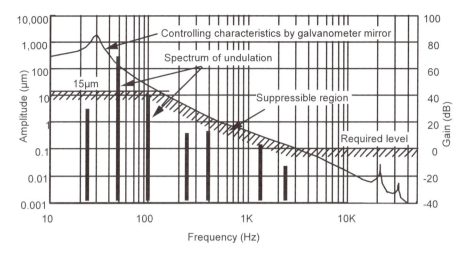

Figure 5.90   Vibration characteristics of optical disk and control characteristics of tracking by galvanometer mirror

Figure 5.91 shows the track vibration. The displacement had a high-frequency component asynchronous with the disk rotation, and a low-frequency component in sync with the rotation. The high-frequency waves had a small amplitude, whereas the low frequency had a large one. The 300μm pp wave was due to the low-frequency vibration.

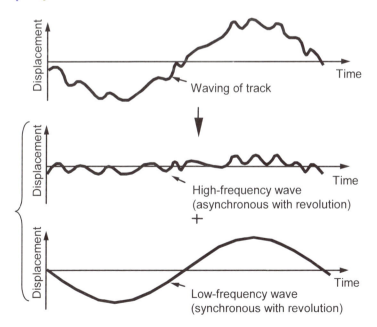

Figure 5.91   Separating the track motion frequency

I then separated the base (Figure 5.92) which mounted the optical system and galvanometer mirror from the base that holds the spindle motor, and moved the former with a linear actuator to control the tracking cooperatively with the galvanometer. I called this arrangement "dual-stage tracking."

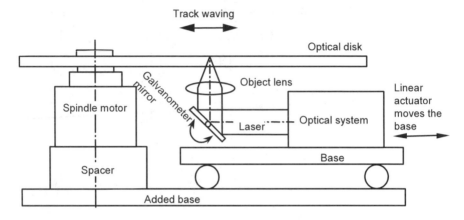

Figure 5.92   Split base for testing the optical disk

A galvanometer is accurate and has a high response frequency but a narrow coverage. The linear actuator, on the other hand, has low accuracy and response frequency but a wide coverage. So I split the tracking signal with high-pass and low-pass filters, as Figure 5.93 shows, feeding the high-frequency signal to the galvanometer, and the low-frequency signal to the linear actuator. In other words, I handled the high-frequency component with small amplitude with the galvanometer, and the large-amplitude, low-freqency component with the linear actuator. Figure 5.94 shows the new tracking control system characteristics. The controllable range was widened and the 300μm pp eccentricity was then in the range.

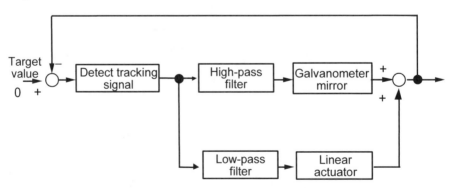

Figure 5.93   Structure of tracking control by dual stage actuator

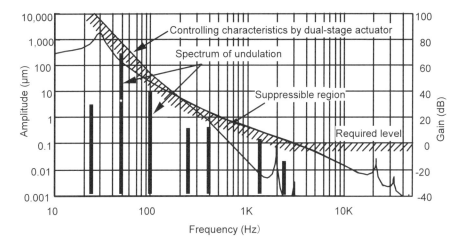

Figure 5.94   Vibration characteristics of optical disk and control characteristics of tracking by dual-stage actuator

Evaluation: I greatly widened the controllable range with the dual-stage actuator. This development eliminated the need to make adjustments every time a get a new optical disk and greatly improved my test efficiency.

Knowledge: A mechanism with high positioning accuracy and a wide coverage can be achieved by combining a mechanism that has high accuracy and narrow coverage with one that has low accuracy but wide coverage. The semiconductor stepper also uses a similar technique: the micro-motion mechanism is provided by a piezoelectric element, and the large motion by a stepping motor with a ball screw.

   Generally, instead of pushing to use a single part with limitation, combining it with a different part that offers the missing feature sometimes works. However, we need to be aware of also the opposite: that if we can make a single part to accomplish a function that used to be realized by two parts, then the single-part configuration will bring the cost down.

5.2 Decision-Making in Technology Management

5.2.1 Selected a 3D CAD System – System Introduction 1

Event: We were about to bring in a high-end, high-priced 3D CAD system for injection-molding die design. Then I saw a demonstration of a mid-range 3D CAD system at a trade show. The demonstration changed my mind and we went with the low-price, mid-range 3D CAD system.

Background: High-end 3D CAD at the time cost about US$30 to 50K for the hardware and US$50K for the software, a total of about US$100K for each workstation. Midrange 3D CAD then lacked satisfactory performance, but it cost about US$10K for both the hardware and software, about 1/5 of the high-end CAD.

I had introduced a high-end 3D CAD for 3D design of molds in 1994 and had been trying to push its use and develop a new way of designing, however, its high price hampered the spread of the system. The design department had licensed another high-end system for the same purpose, which had not spread in the company either.

At that time, newspaper articles ran stories of our competitor companies like "Introduced 300 New 3D CAD Stations" or "New Products Fully Designed with 3D CAD Systems", and we were envious of such companies with deep pockets.

That year, we had budgeted another 3D CAD system for the die design department, but I had doubts if the investment of US$100K would really pay off and wondered if we should continue evaluation with the existing system for the time being.

In addition, the high-end CAD system we had in our department was a different model from the one in the die design department, and another source of hesitation was whether we should purchase another one like ours or switch to a system that was the same as the one in the die design department.

Investigation and hesitation: At the time, I thought mid-range CAD did not have enough functionality and performance and was totally out of my consideration.

I was hesitant with the new purchase however, because the high-end system had not spread in the company as quickly as we had expected. In 1997, I visited a trade show to see the trend in 3D CAD for mold design. As I went by a mid-range CAD booth, I thought "Maybe, I should take a look at the demo," and then my mind went from "A mid-range CAD is not applicable to real design" to "Maybe we can use it."

First of all, the performance was not as bad as I expected. It was not enough for a large-scale product development, but it had enough capability to design molds for injection molding.

Also, it was weak in surface design, which is often used for designing the outside shells of products for appearance; however, products that we designed were primarily functional parts and the weakness in surface functions was probably not too much of a concern (Figure 5.95).

If we could use mid-range CAD for our work, our budget could afford several systems for the price of a single high-end system and we would still be left with some change. I though it would be wasteful to stop using the high-end CAD that I had introduced, but if the work were to start moving we had no choice. Although I was not really certain, I decided to go with the mid-range CAD.

Action: Diverting the budget for the high-end CAD, we purchased two sets of mid-range CAD.

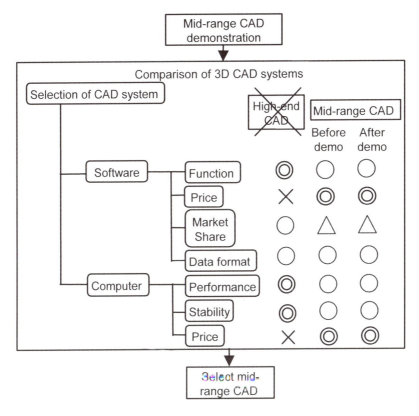

Figure 5.95   Evaluation points for selecting a CAD system

Result: It took half a year to get my hands on the systems, and we spent the remaining half experimenting with the CAD and rapid prototyping. It was then that a group in the product development department planned to bring 3D into their actual work. All of a sudden we got together with the group and developed their actual product in 3D and extended the use of the CAD system into CAM and producing a user manual. We accomplished a great deal.

Knowledge: I was set on the idea that we needed a high-end CAD for designing complex 3D shapes like molds and I almost made the wrong decision. I thought I was up to speed on the advancement of PCs, but the technology had progressed much faster than I thought.

Sequel: The mid-range CAD that I introduced later turned out to be the best-selling CAD software. It has established itself as our company standard and its use is now spreading. I really think that the decision I made back then was a good one.

5.2.2 Arbitrarily Selected a CAM System – System Introduction 2

Event: I selected a 3D CAM system for mold machining from selfish reasons other than functionality or performance.

Background: Curved surfaces were coming into style for industrial design and the need for 3D CAM systems was increasing for mold manufacturing. At the time I was working for our company's manufacturing technology research center and the center already had a 3D CAD/CAM system, but it was too difficult to handle. It was then that the company formed a CAD/CAM group which started its reevaluation of the 3D CAD/CAM system.

The group had a mission to establish the most advanced die design manufacturing model shop for the center. I was also asked to seek orders from other sections of the company. My position was like working for two bosses with different opinions.

Functional requirements and constraints: There are a variety of functional requirements for CAM systems in general. We compared and evaluated (1) functional variety, (2) reliability and (3) operability. We also had to consider which CAM system was likely to receive orders from our internal customers. In that sense, it was obvious that the same system that the customer already had was the best, however, if the system had already been customized to a level at which it could immediately be put to work, then the customization was probably too heavy and it would be difficult to receive orders from other groups.

Mind process and hesitation: At the time, a fully automated CAM system was on the market. It dispensed with the interactive operation of building a CAM model or generating tool paths and automatically generated NC data to a certain level. I will call this system CAM-W. CAM-W was well accepted in the production group, but it had hardly any room for customization or software development and it was easy to imagine that this would make it difficult for us to receive research projects.

Another candidate was CAM-P, which also had automation in mind, but still had emphasis on interactive operation and room for customization. It was, nonetheless, not as impressive as CAM-W.

In terms of functionality, both CAM systems were about the same. The difference was that CAM-W was intended for fully automatic use, whereas CAM-P relied on user interaction (Figure 5.96).

	Operability	Function	Modeling function
CAM-W	O Full automatic	O Required function including everything	Δ None
CAM-P	O Dialogue based operation	O Required function including everything	Δ None

Figure 5.96   Real comparison of CAM-W and CAM-P

If the in-house manufacturing group was our only customer CAM-W that then was easier to use would be most welcome.

The story would be different if other factories (also in the same company) were our customers. In most cases, factories do not like suggestions or recommendations about which system they should use, especially if they have already selected their own system. In this case, the CAM that the factory likes or has already licensed is the best choice.

Course and results: It probably did not make much difference whether we went with CAM-W or CAM-P. The manufacturing group very much wanted us to acquire CAM-W because of their past difficulty with the former 3D CAD/CAM system. We, however, found out that one of the biggest potential customers for contract research was planning for multiple licenses of CAD-P. It was then that I thought that we should go with CAD-P so that we could easily receive contract research orders.

I also thought that if we could find a single case that CAM-W could not handle, then we could claim a reason for not licensing CAM-W. Generally, I knew that most 3D CAMs had difficulty generating cutter paths for a groove with width almost the same as the tool diameter. For molds, this is a typical machining of rib grooves (Figure 5.97).

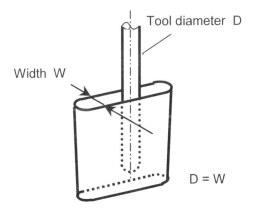

Figure 5.97   3D CAM difficulty with groove diameters close to tool diameter

Fortunately, CAD-W failed in generating the proper cutter path for rib grooves. CAM-P, on the other hand, finished the task properly.

I presented this case vigorously and managed to receive approval for licensing CAM-P. Figure 5.98 shows the table I used then. The table emphasized the single error in rib groove machining as if it was the entire functionality and hardly explained other functions. This may have lacked fairness, but that was the only distinction between CAM-W and CAM-P. Figure 5.99 shows the mind process so far.

	Operationality	Function	Modeling function
CAM-W	○ Full automatic	× Pass failure occurs at rib groove processing	△ Non
CAM-P	○ Dialogue-based operation is fundamental, but batch automatic processing is possible	○ Successfully completes rib groove processing	△ Non

Figure 5.98   Biased comparison chart for selecting CAM-P

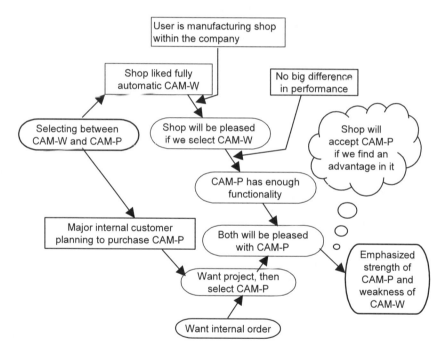

Figure 5.99   Flow of mind process for selecting a CAM system

Afterthoughts: It is probably rare never to have regrets after selecting and licensing a CAD/CAM system. There are many shortcomings first found after starting their use, software patches are frequent and a single system can hardly keep the position for a long time of being the market's first choice. Making a serious comparison under such circumstances is meaningless. We should check if an important function is fatally missing, but if there is little difference we can select what we like.

Evaluation: Users tend to focus on comparing trivial functionalities or how many data file types a CAD/CAM system can read. There is, however, no guarantee that the system will work properly when it comes to production. Instead, we should ask the questions, "What will we use it for?" and "Why do we need a CAM system now?" and then pick the system with the best answers. In this case, the real purpose was to receive contract orders (of course, there were some other factors), and so we matched the customer's software. This reason, however, cannot be the formal one, and so I made an arbitrary selection by searching for a flaw in the other one and emphasizing it.

Knowledge:
- Select CAD/CAM systems on real purpose and preference, not on their functionalities.
- Only check if the functions that are really needed work properly.

> The way to put people in the mood is by:
> (1) Giving them the right to choose (even if things have already been determined).
> (2) Invoking your power or going top-down will only do harm without any good
> - Having them think that they made the choice means no later complaints.
>   If they think they were made to decide one way, a small flaw starts to make everything look wrong.
>   → A good thing starts to look bad.
> - Love blinds a man to all imperfections. But hate a priest, and you will hate his very surplice.

5.2.3 Evaluated a Technology for Breaking Rocks with Electromagnetic Force – Technology Introduction 1

Event: We evaluated a foreign technology for breaking rocks with electromagnetic force. It first appeared like a technological dream and everyone had high expectations, however we did not license it. At the beginning, we only looked at the advantage but later we noticed the shortcomings of having to soak the rocks with oil, with the resulting pollution to the environment and the high cost of removing the oil.

Course: It was about 10 years ago that Mr A, who went abroad to search for new technology, found one in Russia. It worked by soaking a rock with oil, and placing it between electrodes to apply a high voltage. The rock is broken by the electric discharge inside it. Mr A reported his finding to his boss, Mr B who then got excited and enthusiastically explained it the president of company. The president, who was looking for technical innovation, unhesitatingly accepted it without understanding the details. We were asked the study the new technology, assuming that we would license the technology. I had do a lot of thinking.

I had thought about this technology and had run some virtual exercises on it. I therefore knew the benefits as well as its shortcomings, *i.e.*, the drawbacks of using oil. If we tried to soak a rock with oil on a construction site, the oil would leak from cracks in the rock and pollute the environment. The cost of removing the oil from the broken rocks would be horrendous. I thought this technology was wonderful in principle, but required the right place for its application. The company was excited, and we had to try it anyway. So we first conducted a feasibility study. The results were the same as with my virtual exercise and our customer hearing gave the same results as well.

Action: We started the research with first important step of verifying the principle of the technology. We then thought, "How about water if oil is no good?" and even tried soaking rock with water. That was when we found that water worked also, but the power of breaking the rock was a fraction of that with oil, and would have required a huge voltage generator. In the end, we concluded that the method could not be applied to urban areas where there was great concern about environmental protection. It might suit oil excavation or vertical digging, but we not should not apply the principle in developing an excavation machine for general use.

Mechanism of breaking: Our research clarified the mechanism of how rocks and concrete break with electrical current (Figure 5.100).

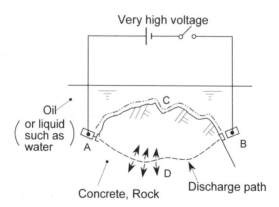

Figure 5.100   Mechanism of breaking concrete or rock with electrical discharge

(1) Press the electrodes A and B against the concrete and apply high voltage to generate electrical discharge within the concrete.
(2) A large current generates through ADB.
(3) The temperature at surface ADB becomes high and the moisture within the concrete turns into high-temperature steam to produce a force (tensile

stress) to separate the concrete vertically at ADB. The concrete within ADBC separates from the rest.

(4) For this mechanism to work, an electrical discharge path has to form within the concrete, and the dielectric constant of oil, which is lower than that of concrete, makes it easy to make this happen.

We conducted our study for concrete, which has a tensile strength that is only a fraction of its compressive strength.

Mind process: Figure 5.101 shows what came to my mind when I was asked to evaluate introducing the technology for breaking rocks in the manner discussed above. First I grasped the advantage and disadvantage of the technology, and then I broke them down into smaller concepts. Then I thought about the obstacles in commercializing the concept, the resources of our company, course of commercialization, and so on.

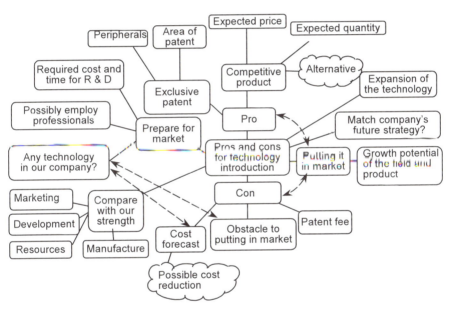

Figure 5.101 Mind process when asked to evaluate unknown technology

The first time around with this mind process, the obstacle to commercialization did not occur to me, but the conclusion that I reached was a natural one. That is, it entered the loop: Pollutes the environment → Avoid pollution and performance drops → Unattractive cost/performance for customer → Small sales → no cost reduction → Unattractive cost/performance … We then judged that it was infeasible for commercialization.

Afterthoughts: Looking back at this entire decision process, matters that often happen with corporate decision-making surfaced. Faced with the competition, once the management makes up its mind, it tends not to listen to the calm expert, and if the top technical person does not understand the principle, he cannot make the right decision at the beginning.

Knowledge:

- Foreign technologies often look better than what they actually are.
- Not knowing the course and just looking at the results can lead to fatal pitfalls when evaluating the introduction of technology.
- Excitement enlarges the advantages and blinds one to the shortcomings.
- If something goes wrong, the disadvantages look larger and we forget the advantages.
- We have to always stay conscious of micro-mechanisms, or otherwise, we cannot make the right judgment when needed.

5.2.4 To Introduce a New Forming Method to Our Main Factory or Not – Practical Solution 1

Event: Whether to push the new technology of forming with lateral vibration for the only molding line of our main factory, despite the opposition from the field, or to abandon my own technology. The manager of the molding department approached me saying, "Unless you abandon it yourself, no one else can stop this project."

Course: The research project started in the spring of 1976 when I was sent to study in the Mechanical Engineering for Production Department at the University of Tokyo. Associate Professor H of the lab that I went to had the following knowledge about packing powder with pressure and vibration (Figure 5.102).

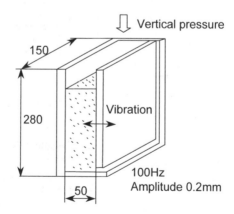

Figure 5.102   Mechanism of horizontal vibration

(1) Vibration is effective when applied in a direction orthogonal to the direction of pressure.

(2) The vibration, instead of vibrating the entire container, is effective when it gives relative displacement to the two facing molds.

My idea was to apply this knowledge to the machine that forms the sand molds for casting. After two years of fundamental study in the university, I proved that this method performed as expected with a packing efficiency much greater than that of conventional methods.

When my years of study were about to finish, the main factory of Company A, which I was working for, had plans to replace its forming machine with the newest model. I returned to the research lab of Company A and as I continued my studies, I targeted the new machine to put my research results into practice. With a gemerous research budget, I built a real-sized test machine (Figure 5.103) and accelerated the research. Within half a year, the research turned into a big project with 15 researchers. Another factor that pushed this project was the support from one of the executives of the company.

Figure 5.103   Actual size of the test equipment

Then within another half year, the project was coming close to be put into practice; however, a concern started to catch our attention. The concern was that the casting frame and model moved relatively, and when the upper and lower molds, which were produced separately, were put together, the two could not fix their positions relative to holes or pins on the casting frame. I had been aware of this problem from early on in stage of the research, however; thought that "If the performance is superior, such a minor issue will be solved quickly." The issue persisted, however, when it came to implementation on a mass production line; and sinceit was the only molding line in our main factory, it was no longer a trivial matter so and the level of concern grew larger each day.

I then came up with the idea of monitoring the two molds with a TV camera to measure their base positions (positioning holes), adjusting the relative positioning with servomotors, and then mating the two molds. We quickly built a real-sized test machine (Figure 5.104).

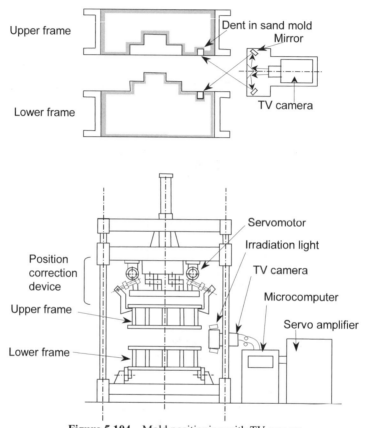

Figure 5.104   Mold positioning with TV camera

The results from this test machine with a new way of mating the molds were, in short, "Will meet the accuracy requirement, but with small confidence level in reliability." At that point, however, we were running out of time for installing the new machine, and the decision of whether it was a go or a no go had now to be made. I, naturally, was convinced that the machine would be put into practice. One day, however, the manager of the molding department, who was also heading the factory, told me, "The top management would not take my recommendation of 'NO' if I said it. I would like you to propose abandoning the technology as the person who has been pushing it."

Hesitation: When we judged from the data we had collected by then, forming with lateral vibration had a better performance than conventional methods, and moreover, there were no distinct data that denied its reliability. If, however, I was asked "Can you really take the responsibility?" then I did not have enough confidence. To take this technology to a success, we had to simultaneously make the two completely new methods of forming with lateral vibration and mold mating with TV camera work perfectly. Not having enough experience at the time, I could not tell what the chances were of the success.

Decision: After tearing my hair out, I decided to abandon my project and requested this to the top management. The decisive factor at the time was not enough confidence in the reliability (Figure 5.105). As a result, my research of forming with lateral vibration suddenly came to an end.

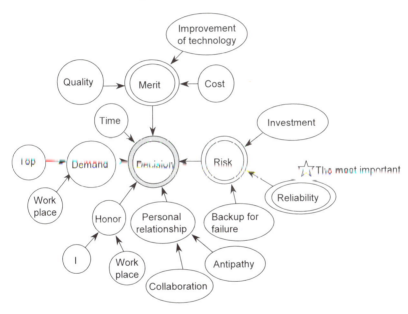

Figure 5.105   Elements that impacted the decision-making

Evaluation: The decision was a tough one for me because I had little experience, and not enough knowledge. It is now about twenty years since that time, and today as an experienced engineer I can judge with a high level of confidence that "If I had introduced that technology into the line it probably would have caused a lot of troubles for production."

Knowledge: People make judgments based on their knowledge, which includes experience. Whether someone has the ability to decide or not depends on whether he has the proper knowledge or not. If one has to make a decision in the extreme state, "personality" and "courage" do matter maybe more than "knowledge". In most cases, however, knowledge matters more.

If one has biased experience, he makes the wrong judgment. Sayings like,
"Every success is a stepping stone to failure" or
"A scalded cat fears cold water"
explain how misinterpreted experience by individuals can lead to misjudgment later.

Sequel: One day, 15 years after I had to put the lid on my project, the president of a casting machine company came to visit me saying, "We would like to put forming with lateral vibration into practice. Although the patent has expired, I wanted to pay a visit to the author of the paper."

I had been away from casting machines for about 15 years and I talked with him about forming machine technology during that time. According to him:

- Research results and reports we made at the time were telling the truth.
- No significant breakthrough had been made in terms of forming machines and forming with lateral vibration had not lost its shine as "state of the art" technology.

The conversation gave me a heart-warming feelings almost as if a child that I thought had died was alive and well.

6

Decisions about Individuals and Organizations

This chapter lists decisions made about individuals and organizations. Section 6.1 is about occupations and 6.2 about decisions for operating organizations. As listed in the Table of Contents, all articles are sequentially numbered and have the topic area titles on the right.

Section 6.1 lists topics in the following three areas.
(1) "Job Change" articles are about jumping from one job to another, or starting a new job after retirement.
(2) "Entrepreneurship" articles are about starting new businesses.
(3) "Turning" are stories about of people making up their minds about new courses at certain points in their lives.

I have categorized articles about corporate management into the following four topics.
(4) "Operation" articles are about decisions the heads of organizations, *e.g.*, corporate presidents, division managers, university professors, made to direct the course of organizations.
(5) "Investment" examples are about capital investments made to change corporate operations.
(6) "Resources" articles are about human resources the most important factor in determining the power of organizations.
(7) "Management" articles give some examples of managing organizations.

As in the previous chapter, articles in this one are written in the first person. Although the topics seem to be a collection of unrelated events, reading through the entire chapter will give insights into decisions about individuals and organizations that relate to production.

6.1 Decisions about Occupations

6.1.1 Jumped to a Small Company for a New World – Job Change 1

Event: When I was a middle player in our company, I started to hate being buried in a large corporation and changed my job to work for a small company. My work then started to affect the entire company and I also recognized the pros and cons of working for large companies.

Background: After getting my master's degree, I worked in the R&D department of Company O. I was enthusiastic about my work which involved one of the leading-edge technologies, unlike many other routine engineering jobs. The work gave me the opportunity to advance my technical skills and I was satisfied with the arrangement.

Course: When I was aged about 30, however, I started to think about my life plan, like when do I buy a house, and I then began to feel anxiety and dissatisfaction; for exaple:

- My work involved one of the leading-edge technologies, but all it did was improve existing products, without any thought to new businesses (I had no impact on the company).
- The company was hardly growing, and yet it had a large number of bright engineers in about the same age range (there were concerns about my future position, and income).
- There were no new businesses and equal evaluations (no distinguishing factor).
- I wanted to move upward to influence the entire business without just staying in the engineering group, but such a move seemed impossible.

I then started to think, "The company size does not matter; I want to be involved in a job that affects the course of the company," and I started to consider small companies. I thought they would have the following advantages (Figure 6.1).

- I wouldl have more influence on the company.
- I could see my work from a higher position.
- I would overlook the entire business instead of just from the engineering aspect.
- Smaller organizations are better for innovation.

I also evaluated the risks for making the move.

- My salary would go down → This should not matter if I scored higher. (I also thought about moving out of the urban area to lower the cost of living.)
- Smaller companies would have their own problems → Well, I should be able to handle them.

The figure shows what came to my mind and important factors are shown within double lines.

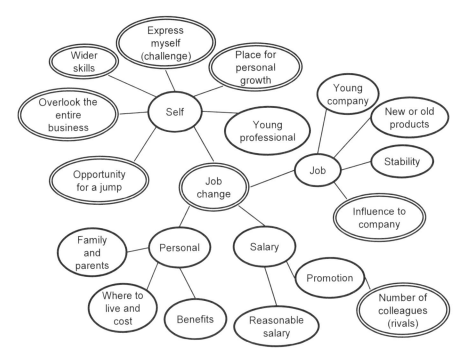

Figure 6.1   Relation of thoughts diagram for deciding to change my job

After all, I wanted to do something bigger. Having no mortgage to pay off, I did not think too much and made up my mind within two months from when I started to feel dissatisfied. When I talked to my wife, she agreed with a smile and I then started looking for a new place to work. I intensively searched the Internet, but the trigger was the calendar Company S put out that hung in the restroom in my home. Once I had purchased many parts for running my student projects from them, but at the time I was not really interested in them. A sales rep from the company had given us the calendar the previous year. The calendar showed their engineering center in Ishikawa prefecture, and it interested me because I was thinking about lowering the cost of living. My wife's parents' home was there and I had some association with the area as well.

As my research progressed, I found that my specialty matched the company and its size was small with 250 employees. The company, moreover, placed an emphasis on new technologies. I started to really get interested and after reading their recruiting ad for hiring people that had already worked elsewhere, I applied for a job. At the interview, the company offered me a position with good conditions and I made up my mind then. This was only about a month after I decided to change my job.

After I made the change, I was bewildered about the difference between large corporations and small companies. However, I am now in charge of product development for a group whose sales account for over half the company's revenue, and I am starting to feel satisfaction with my job. The range of works that I carry out is no longer limited to technical and product development, and I am busy with new product planning, presentations to the top management, manufacturing technology, engineering management, quality control, intellectual property, customer service, special-order processing, and negotiation with the sales and manufacturing departments. The variety of all these tasks keeps me conscious of how I can lead our business to success and I am extremely satisfied with the jump I made.

Afterthoughts:

(1) My concept of "occupation" underwent a big change. The recent long years of recession have made the future unclear, even if you work for a large corporation, and the worship of large corporations and lifetime employment is losing ground. In particular, the younger generation is more flexible in terms of occupation and many are jumping companies in search of their own possibilities instead of working for just one for a lifetime. Large corporations, on the other hand, even without much growth, house a large number of good engineers and their worries about future position and salary are starting to surface.

(2) Figure 6.2 is my view of the differences between large corporations and small companies. Small companies have a number of people with personalities that are unseen in large corporations. Those in the higher ranks may lack broad knowledge or technical skills, but they are energetic, good at negotiations, full of guts and fearless in directing the business. I think the confidence that their work has a great effect on the business has a lot to do with their performance. Also, each and every employee stands out in the company and thus they tend to work harder than those in large corporations. Large corporations keep everyone at the same level, but small companies have to pick out young but capable people to keep the energy level within them high.

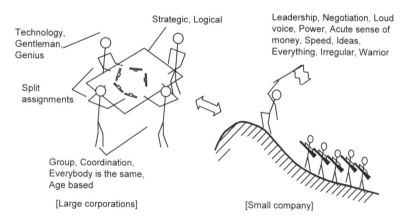

Figure 6.2   Difference between large corporations and small companies

(3) Companies need technology and leadership. Large corporations have good educational systems and accumulations of various technologies, thus, one can pick them up at early ages. On the other hand, with small companies, employees are given responsibilities early on and they are quick in zforming acquiring leadership. Figure 6.3 shows how the qualification requirements of engineers change as they build experience within companies.

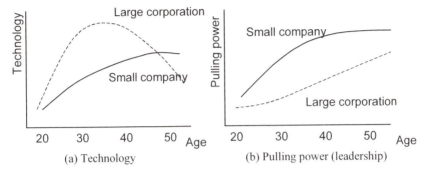

(a) Technology   (b) Pulling power (leadership)

Figure 6.3   Change in qualifications required to engineers

Knowledge:

- The concept of occupation is rapidly changing among young engineers.
- Large corporations excel in terms of early technical education (3 to 5 years).
- Planning a job change is difficult. It needs to wait for the person to change his ideas.
- Making a move is generally better than wondering (but this is not always the case).
- Coming across a good company is a matter of chances or is it fate?

6.1.2 Moved from Manufacturing to Consulting – Job Change 2

Event: In July of 1999, I resigned from Company A and moved to Company B, which closed down 2 weeks later. Within 6 months, I then moved to Company C.

Background: After graduating from the University of Tokyo with a master's degree in March 1993, I started working for Company A, and was assigned to its subsidiary in April 1993 in the R&D and manufacturing department. The later collapse of the bubble economy caused the factory to close down and I was reassigned to another subsidiary and worked there for 4 years in the procurement department. The work was not bad, because it required a wide knowledge of the field for the microscopic phenomena, however I started to think about switching my job because I wanted to work in engineering.

Then I ended up jumping a company; Figure 6.4 shows what came to my mind at the time.

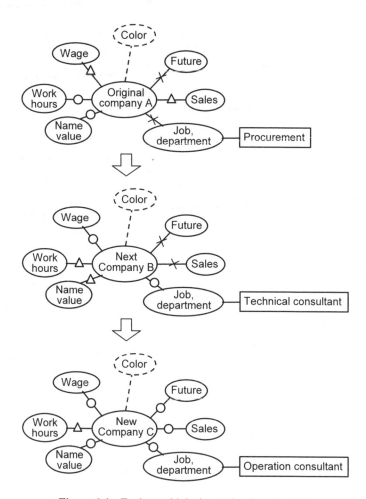

Figure 6.4   Topics to think about when jumping jobs

Functional requirements and constraints: The original functional requirement (purpose) for changing the workplace was to "get a job in the technical field" instead working in the procurement department.

My previous job resulted in a failure experience; *i.e.*, not having a place to return to because of working in a new business outside the company's mainstream at a remote location. I therefore set the constraint of "work in the mainstream business of a company." I also set some other constraints, including "earn good wages" and "secure some time to spend with my family."

Mind process and hesitation: The shutdown of the office soon after I started working there gave me the desire to gain core competence in a technical field. Also, society had changed and working for a large corporation no longer meant a stable job until retirement. I then thought about what I wanted to be doing 5 years later and I set the career path to reach that point.

I also considered staying in the same company but moving to another group, but the R&D department was under a number of reviews and I started to look outside the company. When I analyzed the option of overseas study for MMM (Master of Manufacturing Management, MBA), it was just putting off the decision because of not knowing what to do then and hoping overseas study would provide the next step. I realized that going overseas had turned into the purpose and decided to revisit the option again, 5 years later. I also put off the alternative of starting up a new venture business until my children were in high school and had less time to spend with their father. In fact, starting up a new business would require a lot of work and I would have to work day and night without any days off. I had no urge for a specific business that I badly wanted to pursue or to avoid the stress of working for someone else.

For all the above reasons, switching to another manufacturer was the most inviting choice. I had as many interviews as possible, including with several foreign manufacturers. Through all these interviews, I came to realize that I wanted to be involved in business planning. Machine design was a process of fulfilling a requirement with a machine but I wanted to work in something more upstream like deciding what requirement we wanted to meet or whether a new product was really necessary.

One day, when I was pondering such thoughts, I found an ad in *Nikkei* newspaper for an open position at the parent company of Company B for creating new technology-based businesses, and transferring such R&D outcomes to companies. I thought "That's it," and immediately filed an application. When the president of the company interviewed me he said, "I think what you want to do best fits working with Company B. I am also the president of Company B, so let's have you join B." The parent company of Company B was a US-based non-profit research organization, one of the largest in the world, and Company B was a subsidiary in the technical strategy consulting business

Decision: I decided to work for Company B.

Reasons: The reasons for the decision were as follows:
- My new position met one of the first requirements, of working in the technical field.
- The work I wanted to pursue would keep me in the mainstream of the company if it was a consulting company in business planning.
- It was likely to meet the requirement of higher wages compared to regular manufacturing companies.
- Few people knew about the company and it met the functional requirement that came to my mind later of wanting to do something that others do not.
- A year and a half had passed since I had first talked about the career change. I was getting tired of being asked if I had made the move whenever I met graduates from my university lab.

Action and results: Two weeks into my working for Company B, we received letter from the president in the US headquarters saying that they were closing the single consulting department (where I belonged). Half a year later, I started working for Company C which was in business consultation and system construction. The requirement of "work in the technical field" was my block; I think what I really meant was the higher-level requirement of "satisfy my intellectual curiosity," and I am still working here in management consulting.

Discussions: Unlike my first decision when I graduated from school, I researched the company, did a lot of thinking and even asked many people for advice. The new company, nevertheless, closed even faster than the first had. I quickly realized that I had neglected to look at the company's future prospects, or its current sales. After looking into consulting companies, I settled down with my current job.

Knowledge:

- It is hard to predict the future, but you need to run virtual exercises beforehand or else you will have regrets after making the decision.
- People cannot make decisions without positive thinking.
- Once you have made up your mind, think positively about the results. No one can tell what may turn into a fortune.

6.1.3 Moved from a Company to School for a Diploma – Job Change 3

Event: After graduating from college, I was working in a research lab at an electrical company. Then one day, the professor of the lab I graduated from asked if I wanted to come back and I went back to school as an assistant to earn my degree.

Background: It had been about 8 years since I graduated from the university and worked in a research lab at an electrical company. I had gone through a number of tasks and was getting used to the company, but at the same time I was building frustration in myself. I wanted to write a doctoral thesis, but the busy daily work kept me from writing it. Then one day, the professor of the lab I graduated from asked me if I wanted to become an assistant.

At the time I had the following frustrations (Figure 6.5). The parentheses () in the list that follows the figure enclose afterthoughts.

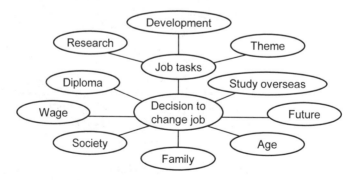

Figure 6.5 Things that came to my mind when I jumped from a company to university

- Wanted to write my doctoral thesis: I wanted to write a doctoral thesis. The work at the company did not quite meet the requirements for a doctorate and I tried to write one that extended my studies at school, but my company work kept me from doing so.
- Project change in 1–2 years: Hardware and software, basic and applied – my projects kept changing rapidly. I had lost track of what I could claim as my career. (Later learned that I was being put through the "manager education" course to gain some experience in all departments.)
- Wanted to conduct research: I wanted to conduct basic research, but most projects that were given to me were about product development. (Research only in a private company is not really a good choice. At least you cannot make it to the top.)
- Wanted to study abroad: I was the first among everyone in the same year that passed the internal English exam for overseas study, but I was never given a chance to go. (The company could not afford to send someone just for the purpose of overseas study. They needed a reason from the business standpoint.)
- Moved to a factory: I was sent for a two-year assignment to a factory, but because I was in charge of technology development, I was making frequent trips back to the lab. I could not understand why I was sent to the factory, which was in a remote part of the country. I wanted to go back to the city.

Mind process and hesitation: Of course, from the standpoint of wanting to write a thesis and conducting research, the university offered a much better environment; however, it was not free of concerns.

- Age: I was about 34 then and it was the last chance to make any career move. The position offered by the school was an assistant and I was a little old for it.
- Salary: Salaries at my company back then were scheduled to go up rapidly after the age of 35. It would be a pity to leave the company then. However, I was still single at the time, and I could commute from my parents' home if I went back to the university. Even if my salary went down, I could probably manage it.
- Future: There was no promise for the future even if I went back to school. In the worst case, I could probably find a new job. The factory assignment was for 2 years, but there was no guarantee that I could be back in the lab after that. I was not comfortable with the atmosphere in the factory either.

After pondering about my choice, I decided to go back to the school. The biggest reason was the thought, "I will later regret it at least 10 times in my life if I do not go back now." I thought regretting without making the move was the worst choice.

Course and results: I received the offer in early April, pondered about the selection for a few days and gave my acceptance. I told my boss at the time in August and moved back to school in December.

Many of my colleagues probably suffered the same frustration, and before I left they often said, "You are going back to the university, I hear. Sounds wonderful. I really envy you."

My doctoral thesis took me three years after changing my job. If I was still at the company, it would have been difficult to finish it. I was later promoted from assistant to associate professor, my current job.

Evaluation: Thinking about the job change now, I think I made the right move. However, had any part of the situation back then – job, location, age, family, or economic situation – been different, I might not have been able to make the change.

Knowledge:

- Action first: This might seem an irresponsible statement, however, a change often leads to better results.
- A goal is everywhere: Wherever we are, we can make something. But we may find another mountain that is higher. We never know until we actually climb it.
- No one knows what will happen tomorrow: The situation that surrounds you may change all of a sudden. The electrical industry went into a long recession and colleagues from my previous job, when I meet them, say, "You really left at the best time." But 10 years from now, they may be in full bloom.
- Not until you die: In the end, I do not think we can judge whether a certain choice at a certain time was good or bad until we die.

6.1.4 Resigned from a Trading Company and Became Independent – Entrepreneurship 1

Event: Left a mid-sized trading company I had been with for 20 years, and started my own business.

Background: After graduating from the mechanical engineering department of a university, I started working for a mid-sized trading company. Most of my friends from university went for manufacturing industry, and the few that went for trading companies selected large ones. I picked the mid–sized trading company because I did not like doing the same thing as other people, and thought I would gain more for working for a smaller one.

Just like any other small company (although it was listed in Section 1 of the Tokyo Stock Exchange), the company I started working for was a one-man operation by the president. The company was profitable, the president held all the human-resource power, and no one could really go against him.

In such companies, the president is everything and all the employees look only to the president for their work. They have learned from experience that anyone that talks differently or blames the president for mistakes will lose out and that the best job security is to keep flattering the president.

The president, no matter what he says or does, is never held to account and wherever he goes is treated with dignity, and that is when he starts to lose touch with reality, be proud of everything he does, and think he is almighty. The flatterers never let him hear bad news, and the president eventually loses the capacity for jjudgment. Everyone in the company I worked for was deeply immersed in the "one man and flatterers syndrome."

When I was 35, I established a new business within the company of building computer systems and became the person in charge. Because it was a subsidiary, I had to pass my proposal for development and capital investment through the parent company. Nothing seemed to move forward at the time.

When a flatterer had to decide about something that seemed to have a low success rate, or something he did not understand, he first thought about securing his own position. They would never take any risk, and would always insure themselves by asking the question, "Are you sure that the project is profitable?" With such an attitude, the company could never venture into something new or innovative. Running the subsidiary company was stressful and gave no opportunity for growth.

At the time in Japan, the money game was the trend and the one-man president was promoting it in the company. A gorgeous dealing room was built, and of course no one could voice any opposition.

In the beginning, the money game went well but soon the big-yen stories of profits that used to leak out from the accounting department stopped.

The financial work is a highly competitive zero-sum (minus-sum for the public, who always have to pay the banker's fee) world crowded with all the professionals from around the world. Without ample capital, correct information, and knowing how to play the game, anyone would easily meet catastrophe in it. And our president had none of the above. (Later, I overheard a frequent visitor, a salesperson from a bank confessing, "I thought the president was going to wreck this company." I was not surprised at all.)

My subsidiary company had its own difficult times, but started to look promising with a profit of approximately US$200K. The president, in a bonus assessment meeting, asked, "Is that all, with 20 people? I would have made a profit of at least US$100M if I had 20 people." That is when I realized my efforts would never be rewarded in the company. In fact, those who flawlessly carried out routine daily work got better evaluation and were promoted quickly compared with those who had to struggle with new businesses. I strongly felt that the company that never challenged anything would be going down soon.

I thought the company would be going down and I was tired of reporting to the flatterers. My hard work took away my weekends, and one day my vision went black and I slumped into my chair. (It probably was the first symptom of my diabetes.) I thought I should leave the company. Fortunately my subordinates had built up enough experience and I thought they could carry on even if I left.

I did not tell my decision to anyone in the company until the very end. I did not even tell my family, who had opposed my last attempt to leave when I was 35. I informed them after I left the company.

Functional requirements and constraints: Although I had decided to quit my job, I had not decided what to do. I was already over the edge of 40 and thought finding a new job would be impossible. Because I was a salaried businessman, I had the mortgage payment but no savings. I had not even a qualification to help me start a business.

70 to 80% of people who start on their own fail. I made up my mind not to bother other people, even if I failed in my venture. That is, I would not start something that required a large capital and I would not borrow. Also, hiring others would cause much headache and I set my mind on something that would not need many people. Another constraint was to assure enough income so my family could maintain the same standard of living.

I think that ages 35, 42, 49, 56, 63, and 70 are turning points for many people. They come at 7-year intervals. Establishing the subsidiary was at age 35, turning independent 42, and at 49, I made up my mind set my mind to go back to graduate school in law and I passed the certified tax accountant exam.

I resigned from the company a month before my forty-third birthday, and established my own company. Any business will require at least 10 years before it really is stable and it was my last stage to independence.

Mind process and hesitation: With the harsh constraints, I could not decide what to do. Supporting my former professor was one possible job, but it was not enough for even a moderate life. I took a vacation abroad for a month, and then started to take action.

The reasons for leaving the company were: it did not suit me, I had lost faith in the company and my future with it, I thought that with the long working hours I could make the same amount of money anywhere, I did not want to work for the flatterers, and at my age it was my last chance of becoming independent.

I ran a number of virtual exercises and had to think hard before I decided to leave the company.

First was the worst-case scenario. "What happens if I fail to be independent?" "I would look bad," "Do I need to kill myself?" "No, I don't have to go that far if I am free of debt." "I may have to prepare myself for a divorce" and such and such. I concluded that I could prepare myself to be a cab driver as the last resort. I remember this whenever I ride a cab and I always leave a tip of a few dollars.

What was the best-case scenario? I was unwilling to take a big risk and did not have to bother even thinking about this.

Without qualification, or capital, the only way to let myself feel easy was the extreme optimism that "If I can't make a living, no one can."

I had left my job and I had nothing but deep water behind me. In fact, it was more like I had already fallen in the water and it was up to my chest. During the first 5 years of independence, I could have easily drowned myself with the slightest wrong move.

I was determined to work to bring benefit to the people, and never indulge myself in a business that woud make me look ugly. I also decided that I had to work three times longer than regular people. I had already made up my mind that if I could not afford to live that way, I would live by my favorite lifestyle of driving a car. During this time, I did a lot of thinking but never hesitated. Thinking brings useful things but worrying does not.

I also thought about applying the "rule of 2 and 8." It means that anything in this world is governed by 20% of the resources covering 80% of the outcome. If it was correct, I could make 80% of my income with 20% of my effort. Then if I worked three jobs, with a total of 60% effort, I could make 2.4 times my assumed income. Of course, things do not go so easily but having three jobs would give me insurance for stability in case something went wrong. When the world is moving fast, things happen like the Japanese text typists suddenly losing their jobs. In any case, I had decided to be independent by holding three jobs simultaneously.

I was not sure what to do for 3 months after leaving my former job, but it was then that my former teacher gave me the job of managing an office building and a friend put me in charge of a new company that built parking lots. After turning and twisting I took off with two new jobs.

Course and results: I became independent in 1991, when I thought the economy was going to pick up. Despite my expectations, it never did. It was the beginning of the "lost 10 years" after the pop of the bubble economy. Those who started before the bubble economy experienced rapid growth, but because of this very growth, most of them saw bankruptcy during these 10 years.

Under the difficult economic situation, I moved my office from the front to the back row, and even quickly moved it into my house. At the time I started an insurance broker age and my friends from the university lab helped me out. I was really thankful for having good friends. After all, I could not afford a bankruptcy and could not let my family starve.

In 1995, however, because "The world is as kind as it is cruel," I got good orders for cleaning up the mess left by the bubble economy. The job was a consultation for winding up the affairs of firms with excessive debts and real estate. After that, work started to roll in, and my income reached the level of regular businessmen at around the same age.

It was then I had to pay for working three times as long as regular people and not caring about my family. I worked so hard to keep the business running and sometimes had to spend the night in the company. My family were able to afford a reasonable life, however the heart had flown away. In particular, the gap between me and my then wife was no longer repairable. Our two sons had already grown up and we decided file a divorce. We settled with her getting ownership of the house. It was the only property we had and the settlement seemed to have minimized the cost.

Evaluation: After turning independent, I had to prepare everything by myself. All the responsibilities just come back to me and so I had to learn everything, unlike a business person who can rely on the company name for receiving orders, borrowing money from the bank, and getting help from other departments. It was my own reliability, judgment, thinking power, management, application, sales, and so on that I could base my responsibilities and judgment on. Through my self-employment, my ways of thinking and looking at matters turned multi-facetted and complex. The start of my own business gave me hardships during one time in partcular, but overall it made me grow bigger as a human being and it was overall a good move.

Knowledge: When you turn independent, you need a mentor. In my case, my former teacher gave me jobs, also advice and encouragement. He did not spare me when he scolded me, but his insights were correct and I valued his advice as one from "Scolding litmus paper."

It is also important to have friends who you can confide in and you can make such friends only when you are students; when there is no conflict of interests with each other. Once you are in society, such friends are hard to come by.

It is people that bring work to you, and so networking is an important aspect.

A human is not all that strong. You need someone who will listen to thoughts that you cannot divulge elsewhere. A counselor is a professional that can help you but it is better if you have someone near you that can meet such a need. I now have a new person that can help me through difficult situations and I have started a new life.

Being independent means all the responsibilities come back to you, and it is difficult to gain a sufficient amount of work. If the situation allows, you need to acquire a strong public qualification for security.

Be warned that a divorce after you are middle-aged may cost you everything you have.

At turning points in your life, *e.g.*, divorce, job change, independence, projects in new areas, people tend to think much, worry, and plan all the details, however, with everything changing so rapidly, most of the thinking is wasted.

Only think about the important factors and make plans; they are all that is needed and they are sufficient.

Figure 6.6 shows what to think about at the time of divorce, Figure 6.7 for changing jobs, and Figure 6.8 what you need for starting a new business.

Additional Note: If we look into people's ability to acquire something new and to have other people think the way you want them to, they cross each other at the age of 38. That is the age limit for jumping into a new field.

Figure 6.6   Things to consider upon divorce

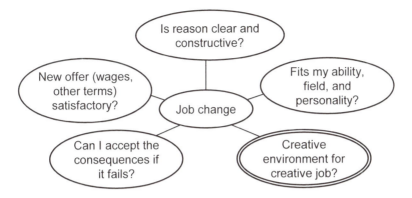

Figure 6.7   Things to think about for career transfer

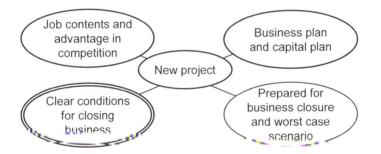

Figure 6.8   Topics to think about for starting a new business

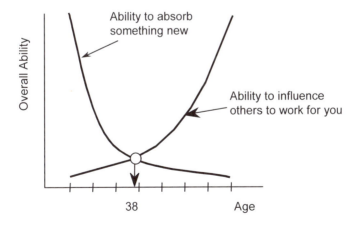

Figure 6.9   Relation of age and abilities (age limit to switching to a new job)

6.1.5 Started a New Business in the US for Building a Bridge over the Pacific – Entrepreneurship 2

Event: In October 1999, I put an end to making a living by working for someone else, turned independent and started a new business. In July of the following year, my friends invested about US$100K in my business to form a limited partnership. Although it has been hard making ends meet, it is steadily growing. The business is about technical translation and information processing.

Background: The time was 1999, when the IT bubble in the US was starting to go down (it was only later that we said so). Japan, which seems to always follow in the footsteps of the US, had news of new companies with a single share priced at several US$100K. People only talked about someone hitting the jackpot, or driving around in a Ferrari.

The company I worked for at the time was a trading company dealing in low-tech parts for computers. With supply chain management gaining much attention then, the high-quality, high-price products the company had were showing a downward sales curve.

Personally I had been divorced for three years, and while I had been paying the child support for my son, I was at a stable time in my life. Being single at middle age was exciting. I could afford to venture out once in a while, but maybe something was missing in my life.

Functional requirements and constraints: There were no functional requirements from the outside. All the investors said, "This money is yours. You can use it as you wish, and I have no intention of talking about your business. I think it will be lucky if there is a return on investment in the future." So I set the functional requirement myself. It was to "build a bridge between the US and Japan."

The constraint was not to let the company go bankrupt. In other words, not to let the accumulated loss go over US$100K.

Motivation: Japanese people in international society, because of the education that teaches them not to stand out, and only knowing enough English to pass the entrance exam, could not even speak half the English they wanted to. On the other hand, Americans, who had grown up watching Star Trek, thought they were the world leaders. The Japanese were losing out in the world market for not expressing themselves, while the Americans with their limited perception were losing as well.

Both had problems with the language on the surface, but were paying a premium for information that was not worth the price, were receiving twisted, wrong and embellished information and people and companies were taking losses here, there and everywhere. Even once they had the information, the computer software to manage it was questionable. The same softwares in the Japanese market had incredible high price tags on them. People were paying for them and did not even know how huge was the markup.

A loss in one part is a loss for the whole. I wanted to at least reduce the loss where I could and contribute to building a strong society that could compete in the world market.

Hesitation: To be honest, I had to overcome considerable hesitation. The task was to overcome the worries of "What will I do if it does not go right?" I kept having visions of losing everything and everyone talking behind my back. I had experience in the translation and interpretation business, and even a side business when I became independent. All this, however, was strictly side business. If I worked for a company, I would receive a set amount of salary even when the company was losing money.

Decision: It was then I remembered the words from my former teacher H. In 1983, GE made me an offer of work in the States. I had decided then that I was leaving Japan and we were discussing about our future at his house at a barbecue party. I probably said then that I wanted to become independent some day, such was my fearless youth at the time. My former teacher then said, "When you cannot afford to pay salaries to your employees, you are going to have to commit suicide."

The words, instead of scaring me off, pushed me on. Even today, I am plan to kill myself when I cannot afford to pay the salaries (at least, that is what I think). Then I was no longer afraid of failure because there was a solution if I failed. Killing myself is just a selfish solution and those left behind will have to go through torture, but it is a solution for me and thinking that way gives me the "energy" to carry on at the time of difficulty.

Mind process: I think I have lived a lucky life. My ability to communicate in English comes from having spent my childhood (from 2 to 8) in Australia, which had nothing to do with my own volition. My job search was brief because I had the desire to go abroad and GE came to our department for the first time when I was graduating. Stanford University, where I later spent 4 years of study, was full of entrepreneurship.

The investors have not said a word about the business. One of them is a senior graduate from the lab and he has started his business himself with success. The remaining two are restaurant owners I met in the US. They both are winning against severe competition and are adding new outlets to their businesses.

You cannot, however, start a new business just because you were lucky. You have to think about it well, and if the chances of success exceed a threshold, all that is left is for you to take the leap. I made the jump and at the time pondered upon the following seven aspects (Figure 6.10).

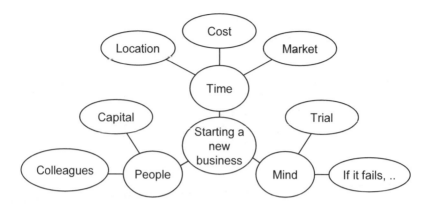

Figure 6.10 Seven concerns in starting a business and three factors for solving them

(1) Capital: At the time, the IT industry had attracted a lot of venture money and one of my friends asked me, "Why don't you try it?" I asked another friend "What do you think?" received a totally unexpected reply: "I would like to invest as well." I thought I had no choice but to do it, and then another person said "Let me be a part of it, too." The sum was much smaller than other big stories that were making the news, but it was enough for the business I wanted to start.

(2) Colleagues: Unless I could find good people, the company would go down. The work of translation does not require much communication with people. It is rather a job for those who tend to stay home. I then looked at Japanese housewives in the US. There were plenty that got married in Japan and made it over to the US. Many had held good positions in Japan and had excellent communication skills in English. That is how they made it here. I started to dig for treasure and made it a condition that the translation work could be done at home. In fact, after I started the business, a number of good people applied for the positions we posted.

(3) Location: The location was Silicon Valley. All of the well-known Japanese companies, and many we had never heard of were here. There is no other place on earth with so many IT-related companies concentrated in one area. The location was the best we could imagine.

(4) Cost: It depended on which city we located ourselves, but registering the company only cost about US$50–200 in Silicon Valley. At the time, getting our home page posted on yahoo.com was free. High-performance PCs and the Internet connection were getting cheaper everyday.

(5) Market: Japan was a huge market for the American semiconductor industry. The Japanese companies looking for sales to US companies were abundant as well. The need for good translation was inevitable. Also, if we introduced the inexpensive service that created the Internet economy into Japan, the country would gain better grounds for international competition. There were two large markets sitting across the Pacific ocean.

(6) Trial: This is always important. You should not just toss around ideas within your head. Listening to people's opinions is often useless as well.

Those with lots of theory but no experience often voice opinions that sound authentic. The success of my side business in translation completed the base of our revenue for the new company.

(7) If . . . : We need to evaluate what to do in the worst-case scenario. I hardly had any burden on my shoulders then and I had no problems living in a cheap apartment for living alone. There was no one to complain even if I was hardly home.

I had answers to all the seven questions. What gave me the answers was people, time, and mind. They would never have come if I was just by myself.

Knowledge: You can never start up a new business unless the following three are lined up.

(1) People: Friends, relatives, acquaintances, and so on. Someone to sympathize with the goal and offer help, people that give you the right market information, and someone that supports your mind when you run into difficulty.

(2) Time: The right time and the situation that surrounds you at the time if it can accept the change.

(3) Mind: After all, it is you that makes the decision. Without a strong mind, it will never come right.

6.1.6   I Became Disabled and Selected the Course of My Life – Turning 1

Event: I became disabled with a problem in my spine, however I can now work and manage my life in a wheelchair.

Background: Two and a half years had passed after I started working for a steel manufacturer, when I started to feel a dull pain at the base of my neck. I thought it was an after-effect from a traffic accident followed by a whiplash about two years before when I had just started with the company. As the days went by, the pain started to grow and I paid a visit to the hospital. After the examination, I was told I had to undergo surgery in the region of my spine and I would be able to leave within 3 to 4 weeks. I entered the operating theatre of a university hospital as a healthy young man, but when I woke up most of my body was numb and I could only move my arms and from my neck up. An unexpected effect occurred in my body and I had become a severely handicapped person.

Course: After the surgery, I lay in the hospital bed all day. The number of tests and treatments gradually diminished as the days went by.

A good friend of mine told me immediately after the surgery that "there had been cases that were cured after 6 weeks." That gave me some hope, however, several months passed by without any sign of recovery.

After about 5 months, I was transferred to another hospital for a test, and the doctor there told me, "You will not recover from the situation and you will have to spend the rest of your life in wheelchairs."

My life then was, except for occasional wandering in a wheelchair, just lying in bed and needing assistance from some other person to do anything.

The days were really hard for me, with the anxiety about my future, and I asked myself, "Is this life really worth it?" (Figure 6.11).

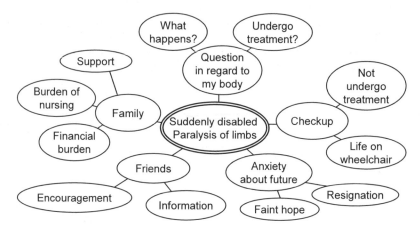

Figure 6.11 Concerns with disability

Action: I then thought that I must seek independence and to come back to society, and I embarked on my rehabilitation. I went through a number of painful treatments, and gradually, through using the remaining bodily functions to the fullest extent, was able to lead an independent social life.

Selection process: After completing the tests at the new hospital, I had to decide which course to take from then on as I would continue the medical treatment. The two options were

(1) Enter a facility with nursing service.

(2) Continue with the rehabilitation.

First of all, my hands and legs were disabled and for orthostatic hypotonia (low blood pressure from standing) caused by the disability, I could not stay sitting in the wheelchair for a long time. I knew returning to my previous job, moving by myself in a wheelchair, and the activities of daily living (ADL) including bathing and using the bathroom were all impossible. That thought led me to think that I should go with option (1) which was at a somewhat remote location and it would relieve my family from the burden of having to care for me.

On the other hand, lying in bed everyday was mental torture. So I also evaluated option (2), to ignore the bodily pain and the small possibility of recovery and target the ADL independence by taking care of all my personal needs from the wheelchair and eventually returning to society.

Making the selection between the two choices was very difficult (Figure 6.12) because, my mind had been totally unprepared for getting disabled, I had had no time to think about the future, the level of disability was severe, and the lack of examples and information made it difficult for me to picture how the future would be.

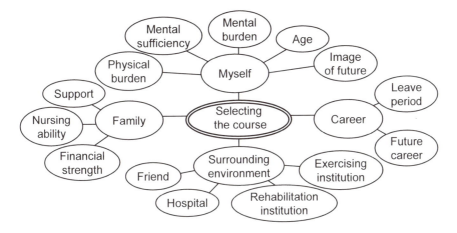

Figure 6.12   Evaluation items for deciding the future course

In the end, thinking about my age, took choice (2), "continue rehabilitation" in order to pursue the possibility of returning to society to the greatest possible extent (Figure 6.13). The following are the factors that affected my decision:

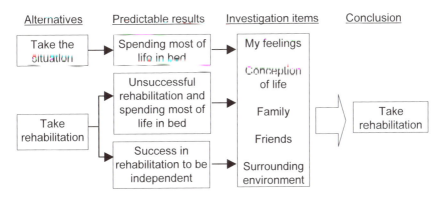

Figure 6.13   Process of making the final decision

- It would be mental torture to stay in bed for the rest of my life, starting in my late 20s.
- My family and friends were supportive of me.
- A national rehabilitation center had been newly built near my parents' house.
- My leave of absence from work had some margin.
- I was not at economic risk.

Results: The medical rehabilitation took over a year. I battled against nausea, fainting, and fever, and gradually gained the ability to move by myself using the manual wheelchair. At the same time I started to cope with orthostatic hypotonia and difficulty in adjusting the body temperature from the disability. These were the results of long hours of scientific therapy, occupational therapy, and sports and my body was starting to make the adjustments needed for the maximum use of the remaining bodily functions (words cannot tell the experience I had during this time).

Within the hospital was an organization for occupational rehabilitation, and after I was released from the hospital, I took the occupational ability development courses there and stayed in the dormitory to build my body power and enhance my skills in ADL. These efforts formed the basis for later being able to work from my home.

All in all, continuing the rehabilitation realized the following aspects:

- Independent ADL and maximum use of remaining body functions.
- Adapting to the wheelchair and performing deskwork.
- Operating a car with manual controls and expanding the range of activities.
- Returning to society by working from home (I have since changed my job and am commuting to work in my car).
- Spiritual satisfaction.

Summary: In my case, in my daily life, I never thought about possibly becoming disabled. In addition, when I started to look for future images of myself with a disability, it was hard to collect information. These factors taught me the importance of a certain level of preparedness in case of an abnormal situation, *i.e.*, where to get the information to base judgments, knowing good consultants (for each field if possible), and having an economic stability just in case.

So far, although I am disabled, my thoughts and the way I feel towards society have not changed much. Of course it is more difficult to physically move around and I often feel inconvenience and bodily fatigue. The way people think, however, when they lead their social lives depends more on spiritual abilities rather than on physical ones. Whether a person feels the life worthwhile clearly depends on factors other than bodily condition.

Knowledge: Here is a summary of my experiences so far.

- If you are prepared you have fewer worries.
- Always ask for a second opinion (from another expert (doctor)).
- The days are hard if you have nothing to do.
- You never know it until you lose it; "ordinary life is also a happiness."
- If you are losing anyway, why not challenge it?
- Even with difficulty (disability), you may make it.
- Personality and ability make a person, not the condition (whatever happens, the way you think does not change much).

Sequel: Several years ago, I got married while I was in my forties. I was rather temperamental before I became disabled, but since it happened, every move I make is slow and I have been forced into being a patient person. I am not sure if that has been something to do with my marriage, but my wife had been my friend for 10 years. The medical term is "acceptance of disability," however, in my case I just got used to it rather than learning it.

6.1.7   Jumped into an Unexplored Research Area – Turning 2

Event: Having completed my doctoral studies, I was then a doctor of engineering. My search for a job found me an assistant position at another university. The lab offered me a research theme that was not in any way related to my past work. I decided to put the gear in neutral and let things proceed in their natural way, and my studies have made great advances.

Background: I finished my doctoral study in Professor H's lab at the University of Tokyo in the spring of 1999. I then chose to start working at Professor E's lab at Tohoku University. Professor E was well known for his work in micro-electro-mechanical systems (MEMS). In the summer of my third year of doctoral studies, I had not settled where I would start working after graduation. In the summer camp (at a beach where we would only swim and drink for 3 days) of our lab, I had a chance to talk with Professor H. He asked me, "Tell me, what kind of work do you want?" and after listening to my explanation he then continued, "Well, then what about working for Professor E?" He then picked up a phone, talked to Professor E, and where I would start was determined that way.

In fact, I was a big fan of cars and motorcycles, and the main reason that I went for higher education in mechanical engineering was because I thought I would find some job in the automobile or motorcycle industry. I believed working for the automobile industry will let me lead a happy life. At school, however, I fell under the spell of Professor H, my advisor, and the sweet words from him and Professor N persuaded me to continue my studies all the way to my doctoral degree.

Course: When I arrived at Professor E's lab, he asked me what study I wanted to pursue. Having completed my doctoral studies in "near-field optical lithography", I had some confidence in the area, but another member of the lab had already started research in the field, and there was no for me to squeeze myself in there.

So I just followed Professor E's suggestion and picked "miniature power sources." The only decision that I made at the time was "Let me follow what Professor E suggests," but later it turned out to be a most important one. Several candidate themes came to my mind then, *e.g.*, near-field optical lithography that I was familiar with, or bio-mechanical research that a young student had started in Professor H's lab. Faced with Professor E's overwhelming knowledge and experience, however, I could not find any reason to stick to my halfway complete ideas.

I set my mind to the area of miniature power sources, and I had to find a theme for my studies. Professor E handed me a folder full of MIT research reports of miniature gas turbine generators. The prototype development for a micro gas turbine element with the size of as a coin was already in progress at MIT using semiconductor micro-fabrication technology. Although an attractive research subject, it was a difficult call for a newcomer because it was already in progress at MIT with big funding and plenty of manpower. So, taking the advice of Professor E to probe one of the most difficult areas, I started research in micro-fabrication for creating a micro gas turbine from a heat-resistant material namely silicon carbide.

"To create something useful" was further advice from Professor E, which led me to look into a micro air turbine that makes an infrared spectrograph polarizer rotate at high velocity.

My next assignment was to get funding for the research. In the year 2000, the New Energy and Industrial Technology Development Organization (NEDO) invited applications for funding by researchers aged 35 or under. It was a great opportunity for me in the first year in research position. I learned about the invitation 3 days before the application deadline, and following Professor E's philosophy of "never turn down something that is coming on its own" I spent the following 2 days completing the application for a funding of US$500K. Fortunately, my application was accepted, but more important than the budget were the lessons I had learned about gathering forces, motivating colleagues in the same project, and having a sense of responsibility as the leader.

The project has progressed quite well, and I can say with confidence that we have made some significant contributions that are presentable to the researchers at MIT.

Another research opportunity I was fortunate enough to receive was a survey of micro-machine centers. Professor E, who could have easily turned down the request, accepted it and gave it to me to widen my specialty. Thanks to the project and Professor E, I enjoyed gaining much knowledge and personal connections by reading through a shelfful of papers.

I gained great confidence by reading about these research activities. My studies concluded that despite the high needs for micro power sources, they tended to be pushed away as future topics due to the extremely challenging technical issues they presented. If there was hardly anyone taking on the topic, maybe I could make it to the top if I worked hard. By then, I knew the needs and what was missing and I set my research topic to work on them.

Once the research results start to lead to the next theme, the mind is activated and new ideas pop up one after another. This activation has kept my mind occupied with the subject of my researches whenever I am awake, even when I am resting my body. By the way, stopping to enjoy oneself at 5 p.m. or spending weekends on hobbies or leisure pursuits and restarting the mind next morning or on Monday are for geniuses. A mediocre person like myself has to keep the mind active at all times, otherwise, good ideas will not reach it. This tendency has sometimes kept me far away even when my wife is talking to me. I am indebted to my wife.

Sequel: Fortunately, the research into miniature power sources is progressing relatively well and has attracted enough interest to obtain cooperation from many people. The number of researchers in the group has grown to over 10 with as many topics. Still occupying the position of lecturer, I am receiving invitations and requests about the topic because it is something I started. A number of people visit me seeking advice. I respond to them positively, in the spirit of "Never turn down something that is coming on its own." Having people understand what I am doing and "selling" my efforts should make any engineering researcher happy.

As I explained above, setting my research area and theme went relatively smoothly, and as I have stated it was the advice from Professor E that pushed me to take the first step. If, at that time, I had insisted on making use of what I had accumulated over my student years, or picked something that just came to my mind, I would probably not have been as happy as I am now. Professor H, when he sent me to Professor E, said, "This guy is the one with most energy (in other words, the most pertinent one) and I would like you to make use of him as the leader for the students." Perhaps, Professor E then decided to prepare an area full of new possibilities but with lots of challenges.

Knowledge (primarily for students with ages similar to mine):
(1) "The frog in the well should learn about the great ocean": If you think you are a frog in a well, following those with knowledge and experience is a good decision by itself. Once you are taught that there is an "ocean" out there, you can venture to it and make new discoveries.
(2) "Let it be": This is Professor E's favorite saying. It means to let the need decide which way you are headed. Let go of your bias and letting society direct you often leads to good progress. Forcing yourself to work on things nobody wants never leads you to success.
(3) "Never turn down something that is coming on its own": When you are asked to perform some task, try to accept it as a service. Pursuing several tasks at the same time and being pushed into a corner often makes the work progress quickly and can broaden your view of your work. If you always keep yourself busy at all time, you will get the positive feedback of receiving another work request. Of course you have to control things so that they do not diverge.

6.1.8   Built a New Lab and Research Group in a Conventional Field – Turning 3

Event: I finished my doctoral studies at the University of Tokyo in 1966, took the job of lecturer in the Department of Mechanical Engineering for Production, then followed it with a position as an associate professor, and moved to the lab for research in machining. It was about 1970 and my research topics at the time involved casting and powder dynamics. The lab was new to me and it made measurements in micrometers, and in intervals of milli-seconds, a world totally different from my previous one of 50% measurement error and a single data point in a day. For a while I was lost not knowing what was needed and what were the unsolved problems.

Five years after I came to the machining lab, I went to MIT on a sabbatical for a year. When I finished my studies abroad and returned to the lab, there were only three black machines sitting there. The previous professor had retired, and I had to restart everything from scratch. Of course, my research funds were almost nil.

Mind Process: I first thought that I should follow the main route of research because that is where I was. I first set four main research areas, turning, milling, drilling, and grinding. Then I decided to learn what industry was looking for through (i) reading an industry newspaper every day, (ii) visiting factories and trade shows, (iii) listening to others, and (iv) attending conferences. The research should (i) be useful to industry, (ii) start from knowledge and technology in my group, (iii) value accomplishments by the group members, (iv) gradually switch to better topics, and (v) avoid waste by building the next year's efforts on top of the previous year's results.

Execution: First, having knowledge of power engineering, I started with research on grindstones. With the research results of the previous lab professor, I asked T (currently a professor at the University of Tokyo), to carry out research in face grinding. This was because we conveniently had a face grinder in the lab.

The first research topic in turning was a cutting tool with sensing capability. The tool itself had the capability to both measure the cutting force and sense the temperature simultaneously. I started this work thinking that we should automatically detect cutting problems and clarify the cutting phenomena.

For the studies of drilling, I found there was a problem of damage to the cutting machine or work when a drill broke during the drilling process. At the time, a fire broke out in an unmanned factory during the night and automatic detection of the drill bit breaking was needed. Such accidents were frequent with small drills of diameter 5mm or less, and I settled on starting research in the area.

In the autumn of 1980, I went to a machining tool trade show with H and he came up with an idea on the way home. First, W made a prototype torque sensor which worked well. Then H started a serious design and, with the help of many people, built a torque sensor shaped as a toolholder for small-diameter drills. The sensor sends out the data via a wireless device and gives an accurate reading of the cutting torque and can predict when a drill bit is about to break. This tool made its way to a machining tool trade show and turned into a hot topic.

I spent 30 years with H as a colleague, as well as a joint researcher. Having him design the hardware and cooperating with him through the researches was my first decision in life, and it was a good one. Good friends are hard to come by.

Academic research requires not only industrial applications, but also the clarifying of phenomena. We then set out to detect not just the torque, but also the linear force and temperature at the cutting tip. We ran experiments of measuring the quantities when the tool opened holes in printed circuit boards. We also obtained theoretical numbers which matched our measurements.

Applying this technique of building sensors inside tools, we built a polishing tool for finishing curved surfaces in 1984. The machine automatically moved a handheld grinder in three orthogonal directions. In 1986, this line of research with some research funds, developed into an intelligent machining center. For this research I went together with H to Company M and explained the functionalities we were looking for. It took the whole day. The machine that came out of this research sent sensor signals from itself to the computer for the calculation of correction signals, and allowed the inputting of the correction with manual pulse signals to the NC machine. At the time, it was epoch-making. Later we recognized that if we fed the signals to a LAN, it would have been the first open-architecture CNC machining center, but we did not think that far and were happy with the outcome as a convenient machine. We took advantage of our development to our next research into milling.

This research was followed with ideas from H for developing a six-axis force-sensing table that measured the three force and three torque components of machining, a fail-safe table that protects the machine and work by running away if it detects excessive cutting force. These developments allowed cutting with a constant force and speed.

When our research efforts were fully engaged in developing the intelligent system, I had one concern. We knew that machining technology was just like other technologies, progressing towards integration IT, but we were limited to engineers in the machining area. I then selected M to be the successor of our lab, he having studied two-legged walking robots in an information-processing lab. At around 1985, the university still had strict rules for managing the missions of labs and so I told him to take the world of machining as far as possible in the direction of information technology, without getting tied up in conventional machining. This decision was the second one in my life that led the lab, the department, and myself in the right direction. Good followers are hard to come by.

M later developed a network-type production system (what is now called e-manufacturing) by integrating an open-architecture CNC machining center, a real time control system, and a workstation, and hooking them up to the Internet.

Later development: In around 1987, we started the development of a high-precision machining center that automatically corrects thermal deformation. This was a machining center that compensated the main shaft position by forcing thermal deformation of the machine column. We mounted the column with a deformation sensor, a cooling device and a heater. The deformation sensor was a strain gauge type and it accurately measured the deformation between two points to 0.1μm. The cooler and the heater corrected this deformation. In addition, the six--axis force-sensing table enabled intelligent machining, the open-architecture CNC controller hooked to the Internet allowed remote operation, and the fail-safe system ran back the table in case of excessive cutting force to protect the machine and work. The machine also detected vibration and modified the cutting conditions to stop the vibration. We demonstrated this machine in the 1990 International Machine Tool Fair.

January 17, 1991 dates the beginning of the Gulf War. A day before, we conducted a tele-operation experiment that remotely operated the intelligent machining center in the University of Tokyo from a lab of Professor K at George Washington University in Washington D.C. It all started from a report by M, half a year earlier, that said, "I successfully operated the machining center in the B1 floor from a computer in the lab on the fifth floor." I then said, "Operating it from the 5th floor is no different from operating it from America," and then had M look into what it took to operate it from the US. We found that the project would cost over US$100K. When we were about to give up the idea, we later learned that running the experiment for 30 minutes would cost only US$30K. After pondering about it for a while, I decided to go for it. I thought it was worth the expense and if I had someone else pay for it, the results would be taken away from me, so I spent all the money I had in my name on the project. The breakdown was 10K for the satellite channel, 10K for borrowing the equipment, and 10K for operators.

When the actual test started, we were surprised at AT&T failing to provide the connection and we had to run the experiment without an image from the US. The test itself succeeded although without an image. M and a student who traveled to the US successfully operated the machining tool sitting in our lab at the University of Tokyo. Later on, a number of tele-operation experiments succeeded among Germany, US, and Japan. It was an epoch-making technology of transferring not only the image and sound information, but also the movement information to a remote location with an intelligent machining tool connected to the Internet.

Many research results came out from H's lab as well, and the industry was interested in them. In April of 1989, we got together and decided to start the Research Group of Intelligent Machining. A number of companies gathered in the group and we reported our results, had them evaluated, and directed our studies in the direction the market wanted. The group lasted for 9 years, during which time it changed its name to the Intelligent System Research Group and then in 1998 continued on to the Creation of Technology Research Group.

In 1993, a budget was approved for Super Machine Creation System which was a prototype factory with a group of intelligent machining tools. It houses intelligent machining centers, intelligent lathes, intelligent ultra-precision face grinding machines, 3D measurement tools, EDM (electron discharge machining) tools, and wire-cutters that are all connected to the Internet. At the time, with the collapse of the bubble economy, universities had a hard time acquiring research funds from private companies and started to obtain research grants in units of millions of dollars from the government. Figure 6.14 shows the process of starting up.

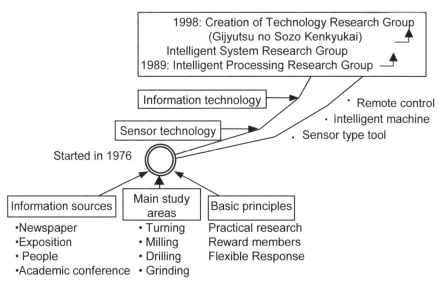

Figure 6.14   History of lab developing into a research group

Knowledge for making decisions: Through my activities, I learned the following aspects that are important in directing researches and making decisions for research organizations. Figure 6.15 shows them graphically.

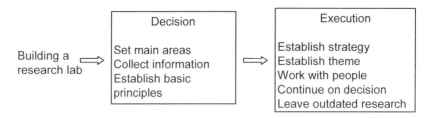

Figure 6.15   Knowledge for researcher's decision-making

(1) Decisions should be made quietly and by yourself. Do not rely on hopeful observations and stick to the possible. Once you make up your mind, stick to it to the end no matter how long it takes.

(2) You will get help from a lot of people. In particular, the of your friends and subordinates determines everything.

6.2   Decisions about Corporate Management

6.2.1 My Company will Disappear in One and a Half Years – Operation 1

Event: I am the president of a subsidiary of a steel company. In 1998, the year when the rapid recession was causing huge chaos, my company was losing sales. In January, although the number of operating days was small, the decline in sales was serious. In May, we faced another decline beyond expectations with a slow month. Although I was puzzled, I had to borrow $US2M in June. The decline in our sales figure continued for several months, and I started to feel the seriousness of the problem.

The media reported that the Japanese economy would continue to go down. In fact, we only heard bad news: the bankruptcy of Hokkaido Takushoku Bank in the spring of 1997, the collapse of Thai Bart in autumn, the bankruptcy of Yamaichi Securities, and so on. At the time, I could not imagine the effect coming so quickly to a small company with annual sales of US$30M, but as I had to watch sales drop over months, I felt the waves of the real recession. In fact, each month showed a loss, from a few US$10K to US$200K. The company had a weak financial standing to begin with and not knowing what to expect, I was descending into chaos.

Mind process in the chaos:
(1) Presuming the worst scenario
    It would lead nowhere if I left the situation. What should I do? The most difficult problem was the finances. So I thought I should forecast the situation and find out how long the finance could stand the continued loss. I was not sure, however, how to project the finances. I went back to the starting point and developed formulas to compute the income and expenses to estimate how the finances would decrease. I set the sales to an independent variable and assumed as follows:

- Income: Assume the worst sales in quantity and price, of what we have now, and add the effect of monthly variation.
- Expense: Based on the assumed quantity sold, compute the cost of production. Add the effect of deferred payment based on the typical payment terms of each customer.
- Funds: This would be easy to calculate because the funds were only the cash on hand, however, most of the income was in promissory notes. The payment term on each note varied from 2 to 5 months. I studied the variation and came up with an average term of 4 months. I simulated the change in funds as they are affected with income and expenses.

(2) Results

Figure 6.16 shows the results of the simulation. The results were astonishing. If the situation continued, the funds would be gone in a year and a half and our promissory note would bounce. I only had a year and a half.

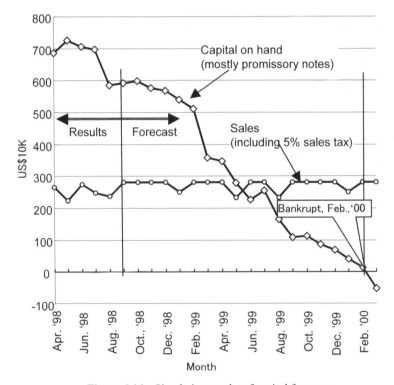

Figure 6.16 Simulation results of capitol forecast

(3) The surprising results pushed me to execute a solution

Given the results, I was pushed to think of a solution and to execute it quickly. After all, the solution was to reduce the operational size to meet the large decline in sales. The time remaining, however, was only a year and a half. Sharing this information was most effective during the difficult operation of the solution, *e.g.*, when asking people to take early retirement. I felt the importance of forecasting the future and extrapolating what to expect.

Knowledge: There are times when it is hard to see the situation around you. In such cases, the following things will give you a guideline.
- List the factors that govern the situation.
- Among those listed, find the most important one and predict how it will change. Base your prediction on actual data and along the real time line.
- Based on the forecast, determine your solution and course of action.
- Once your plan is decided, all you have to do is execute it.

6.2.2 Forced an Organizational Change in a Traditional University – Operation 1

Event: When I moved to work at a university in April 1992, the Department of Engineering Synthesis was not so attractive. The Mechano-Informatics Department moved their specialty to control, information, and intelligence, and the Department of Mechanical Engineering to environment, nanotechnology, and biomechanics. Our Department of Engineering Synthesis still had its specialty in design, industry, and systems, just like it used to in the 1980s. The department had lost interest in its own existence and future. At the time, I was busy with my own research and was not thinking about the department at all. I was busy coordinating joint researches with a number of companies, spreading our group in the way the old Soviet Union used to do. I hardly attended any department meeting, and was receiving a lot of complaints.

Course toward decision: In 1998, the issue was raised of what to do with the department machine shop which occupied over 600 square meters of floor area. In the 1970s the shop had about 10 technicians to produce all the parts needed for the students' experiments. The technicians, however, could not operate the NC machines or perform precision machining, and the department was waiting for their retirement to let the shop disappear naturally. I thought perhaps we could use the shop to revive our department.

Up to that point, every school decision went something like, "We shall assign the area of AA to Professor BB to have him manage the lab for CC, and the lab is well known for its research in DD; let's set the catchword for the department to EE." That was the "bottom-up" way of making our plans. If we applied the method to the problem of the machine shop, we had to find flaws in the current mound of constraints, put everyone's complaint together to form the evaluation function, and solve the problem of minimizing the evaluation. Negotiating terms with those involved was tough (and still is even now). The department was like a community of small shops. The department could not give directions to the professors, each being a shop owner, and matters were discussed until everyone was in agreement. The long discussions that led to no conclusion often caused the department to take actions when it was too late.

Mind process at decision: I had to come up with a plan. If I started to tear the shop apart, there would be complaints. So I needed a justification. If I declared a grand strategy, however, I would have to go through the trouble of doing everything myself. On the other hand, no one else was willing to pick the chestnut out of the fire. If my department went down, my lab could not expect good students and I would have to go down with the department. So I decided to take charge myself, as I wanted to. I started by stating my strategy and discussing the matter in a "top-down" manner to minimize the energy needed for gaining cooperation. Nothing takes longer than laying down the groundwork (fortunately, one of my colleagues was patient enough to go through the groundwork of persuading the other professors).

I then set the catchword for the Department of Engineering Synthesis to "creative design." The reason was simple. It was what I wanted to do. Naturally, some professors complained, *e.g.*, saying that "design" was a target for every engineering department, but I ignored such complaints. Our department called itself the Department of Engineering Synthesis, and if I was asked what "creation" was, I would explain that it meant the high-level design of starting from setting the assignment, *i.e.*, the design that starts from setting the functional requirement and constraints. It is important to "put your thoughts into words, and put the words into shapes," and so as to pull in everyone to execute my own creation, it was absolutely necessary to put the requirements into engineering terms. I said that our country had been putting thoughts into shapes (without the words part) since the Meiji era in targeting western products, and that was the very reason why the "latent knowledge" about creation did not transfer to the next generation. executed the decision, and once it was determined the rest was just a matter of sequence. As Figure 6.17 shows, the department's goal was to "educate leaders for the manufacturing industries", and the method of accomplishing the goal was "have the students experience creative design to generate new industries". The actual plans included to "set creative design exercise as the main goal" and to "set developing new frontiers as the research goals" with the further requirements to "hire a professor with experience in machine design" and to "cooperate with companies with researchers with new design requirements." The target, by the way, for the Department of Mechanical Engineering was to "find the truth in mechanical engineering" and "apply theoretical methods" to accomplish the goal. No wonder we did not get along well (though the two departments later got together and decided to set a common curriculum for creative design).

Sequel after decision: For the class of Exercise in Creative Design, we gave an assignment to students of building Stirling engines. We started by renaming the machine shop the "Mechano Design Studio" and assigned two assistants and two technical officers to operate the studio and to run it by teaching the students instead of selling hourly labour. The students were grouped into teams of four to use manual lathes and milling machines to build a 1W engine that rotated when heated by a gas burner as in Figure 4.51 and 4.52. We encouraged the students to come up with their own design. After they had drawn a schematic, the students each took take turns to use the studio to build the engine within 10 half-day sessions. Other assignments included toys that took sensor signals into the PC to control actuators, designing with a CAD system and building the design with rapid prototyping or CAM/NC. During the winter semester of the junior year, the students were busy building something 4 days a week. The assignment was set and then solved. This gave the first "destruction of the pattern of learning for students who believed that the problem was something that came from the teacher.

Figure 6.17   Plans for the department of engineering synthesis

The new frontiers included cooperation with the medical and bio fields, targeting the collaboration of medicine and engineering, welfare and information fields that targeted meeting individual needs, machining and material fields for producing nano structures of size 100nm or less, and the field of design methodology that makes use of failure knowledge, know-how and ideas for realizing the above. When I talk about my research plans anywhere, other researchers and corporate people gather naturally to share the excitement.

The biggest problem was to hire a teacher with design experience. I had difficulty myself for getting promoted to a full-time professor in March of 2001. I had to promise that I will "act with more decency" to the professorial committee. Once I was promoted, I made this difficult motion of adding a designer to the faculty. I could not tell if hiring a person who had been working in the industry on something like "design" whether he would successfully instruct the students or not. Even if he could teach design, his lectures would lack some punch because they came out of books. The most difficult part in this part of my plan was persuading the other professors. Engineers heavily involved with product design in companies hardly have any time for writing papers, and never enough time for a doctoral thesis. Other professors, nevertheless, after hearing my long explanation would ask, "OK, I understand. And so, how many papers has he written?" At the time all that mattered for hiring was the number of papers. But I should say it *used* to be that way. In April 2002, I hired two engineers with design experience. My enthusiasm had moved the professorial committee.

The evaluation function for my plan was the satisfaction level of students. Averaging the scores from the exercise classes, the score that had been 3.5, after 10 years went up to 4.2. Also, corporations had tripled the research funds and grants from US$1M annually to 3M. In reality, the progress of this project had been "three steps forward, and 2 steps backwards." I had experienced as many failures as successes. The failures, however, did not make me feel bad and they never stressed my mind. If I had spent time and energy in persuading everyone in a bottom-up way to achieve unanimous agreement, I would have lost a lot of research opportunities in addition to failing to hire the new members with industry experiences.

Knowledge:
- If you hesitate about your decision, going in the direction you want to leaves less stress whether you succeed or not.
- A top-down execution of plans, starting from explaining what you want to do, often makes the project proceed smoothly.

6.2.3 Located a New Factory – Investment

Event: One of our old factories was giving problems because it was too old and we were looking for a new location to build a new one. Negotiations to purchase land in one town were in progress, but the existing occupier kept putting off negotiations and the project was delayed by a year. We then started looking for another location, and we found an industrial complex where we did not have to negotiate over the land, and moreover, the geographical condition was favorable. We canceled the first plan and decided to relocate to the industrial complex. We promptly started construction of the factory.

Background: One of our old factories was starting to have aging problems. It was difficult to replace old equipment due to the aging of the entire building, and the factory was losing its competitiveness in the market. Also, the factory might have to face shutdown in case of an increase in local population due to environmental concerns about its metal production line. To add to the situation, the electricity supply from outside was insufficient and the factory could not build its own in-house generator. All these factors contributed to our search for new land on which to build a new factory.

Decision process: Our search for a new location led to one which could host agricultural or industrial buildings. All the landholders agreed to the relocation and the town was going to take care of all the land purchase. We also signed the agreement on factory relocation and pollution control, and we started investigating the actual equipment to go into the factory.

When the land purchase had progressed up to 95%, it suddenly stopped. One landholder resisted selling his land, saying, "I was not involved in the negotiation. Being a full-time farmer is different from others that are old or running the farm as a secondary business. I would like to receive convincing explanations." The landholder did not quite decline to sell the land, nor did he state clearly his conditions for selling. The town hall person in charge of negotiations honestly tried his best to persuade the landholder, who kept coming up with a new condition as soon as one was settled. This frustrating situation lasted for a year.

Meanwhile, equipment selection kept moving on quickly. To avoid spending too much time on the first step of purchasing some land, we thought we could push the negotiation by hinting about the presence of another candidate location to the stubborn landholder. We then went to the prefecture government to look for another piece of land. Unexpectedly, they gave us a big welcome. We had consulted with them about 7 years before, but they had almost ignored us, saying that a metal factory was not high-tech. The sales of land designated for industrial complexes were going down dramatically after the bubble economy had burst. Forced to sell the land, the officers were no longer picky about the type of business we were in. The economic   environment had changed dramatically in the past 7 years.

The consultation had started with the purpose of finding a competitor against the reluctant landholder to push the negotiation in a favorable way, but the industrial complex had excellent conditions. The closest residence was more than a kilometer away. The roads were wide, mountains surrounded the spot, and there was an artificial pond for emergency fire-fighting water. The location, therefore, had no foreseeable possibility of someone coming to live nearby. We had by no means, any intention of polluting the area, but we wanted to avoid unnecessary friction. We did not have to create a green area for emergency water storage inside the factory land. There was a 66kV main transmission line passing near the factory, and we would receive a free connecting service for electricity from the electrical power company. Accessibility to a highway was also excellent. On top of those conditions, the land preparation was complete, so that we could start the construction immediately. In no time, we decided to choose the industrial complex to build the new factory, and took the unsuccessful negotiation back to the drawing board. We had to go through some processes, *e.g.*, reappraise the land price to get approval from the prefecture assembly, attend the pollution assessment and local presentation meetings, and arrange the pollution agreement and building agreement with the locals.

The town officials from the first location agreed to cancel the original plan, and our company agreed to pay some settlement since we caused some trouble to the town officials by holding up other sell-off plans.

Results: The original plan was canceled because of one opposing landholder. As a result, however, we luckily relocated the factory to a prefecture industrial complex with excellent conditions.

Summary: In most cases, agreements from all the landholders are required to purchase land to build a factory, and the agreements are often the biggest obstacle in acquiring land. The problem becomes complicated as time progresses because the number of landholders can increase from inheritance, and in the worst case, may lead to failure in purchasing the land. It is tough to cancel a plan that is in progress, and moreover, the cancellation can cause great damage if the company has already signed contracts with landholders that have agreed.

Additionally, the cost of preparing the infrastructure can be a big constraint in constructing a factory. If a company has to prepare the infrastructure by itself, it has to be ready for unexpected expenses. Land for building factory facilities on private property is governed by such regulations as those dictated by industrial location policy. Industrial complexes built by local public organizations usually have prepared green areas, emergency water reservoirs, and excellent infrastructure, such as roads, water, electricity, and drainage. Figure 6.18 shows the ideas which went through our minds during the factory relocation.

Knowledge:

- When acquiring property to build a factory, target land with less landholders. A local industrial complex or land where another factory already exists is a good proposition.
- "The world is as kind as it is cruel," and to find good luck, you have to keep visiting your target.

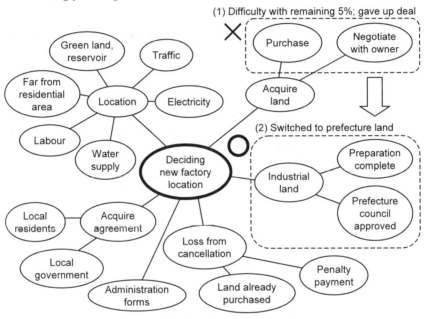

Figure 6.18   Deciding new factory location

Related thoughts: The Japanese administration often purchases land in patches as each landholder agrees with the terms. That is probably what makes it difficult to give up the plan midway. In China, all the land is owned by the government and it can be acquired immediately upon a decision by the government or the governing party. This makes the redevelopment of land into airports, roads, or local public facilities extremely fast compared with Japan where the land is privately owned. The land purchase for Shanghai Pudong International Airport completed in a month. On the other hand, Narita Airport started as a national project 30 years ago and some of the land is still under negotiation. In China faster things are hundreds of times faster.

The choice of whether to purchase or rent is another first selection that you have to make. The tax laws of Japan do not allow listing land as subject to depreciation. A company, therefore, can only pay for the interest to a loan for land purchase our of profit after tax. The cost of renting land can be listed as an expense and thus is an option to evaluate when the time comes.

The compensation for damages when a contract is cancelled can be large for agricultural land on the basis that farming was interrupted even if cultivation followed a normal schedule. It is therefore important to periodically take pictures of the area and keep records of physical evidence just in case. Once a unanimous agreement is reached, you may want to exchange a letter of intent.

Some try to take advantage of others when they see the others at helpless disadvantage. I wonder what the person was thinking when he tried to delay the deal. Also, how he has been getting along with his neighbors since the deal was called off.

6.2.4 Selected My Successor and Recommended Him – Resources 1

Event: Whenever I moved up the ladder in my company, I was asked to recommend a successor for my old position. I gave the matter some contemplation each time so as to recommend the person I thought was right. When I look back at my recommendations now, however, I am not sure if I made the right decisions. For one person to evaluate others is indeed a difficult act.

Course: When I was heading a research lab, the company decided to advance into a new field and I was asked to join the new group. My way of work was not to go into too much detail and once I started thinking about one thing I would be occupied with it all the time. I thought then that for my successor I would select someone who noticed all the details and was in sound health. I thought he would continue with the same research topics and lead the lab to success. When I completed my new assignment and returned to my old post, the researches had continued, but with hardly any new results. The company was disappointed with the outcome of the research and the lab was soon closed down. That was when I realized that someone to manage research had to possess the qualities of producing new ideas and executing them. A good personality was not enough.

At another time, later, I was given the assignment to join a new national project and be in charge. It was an area we had no experience in and I gathered a group of people to form the team. Within a year, we built a new test system and within the next few years used it to develop new products. The success gave me promotion to the next step, and again I had to choose a successor.

This time, I recommended a person who shared the hardships of venturing into a new field. I sensed I was giving him a reward, but my biggest expectation was that he would certainly carry on something that he had participated in starting up. Despite my hopes, the national organization (customer) said that the project started to show lack of new ideas in both its contents and proposals.

I learned from this incident that leaders of research groups should possess the ability to generate new ideas themselves. Come to think of it, I should have known from my first experience to look at people from a different perspective. I regret that I should had grown to look at people's qualifications from a higher point at that time.

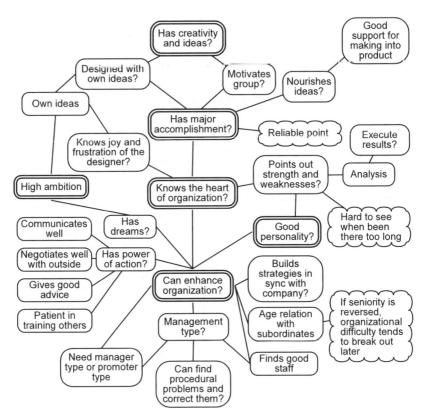

Figure 6.19   Mind activity for selecting the next person

Hesitation: I am now in charge of all the research and development activities for the entire company. It is 30 years since I joined the company. I will be required to choose a successor at some point. How should I base my judgment then? Personality, capability, skills of negotiation, managing organizations, intelligence, or power? I now know that I should not look at only one aspect. I keep wondering what to do.

Summary: Figure 6.19 shows what comes to the mind when finding a successor for a specialized department like a research center. Many aspects enter the mind, and the following are the three important points suggested by my 30 years of experience.

- As a person that lives in an organization, can the candidate enhance the output of the organization, and does he know the basic nature of the organization?
- Does he have the ability to come up with new ideas or methods, and has he made significant accomplishments?
- Does he have a superior personality with high ambitions?

Knowledge: It could be a controversial argument, but I believe someone in charge of a highly specialized position like a research center of a corporation should have the following characteristics. Broadly, they should be someone who has made significant contributions, is of the right age, and has a good personality, but he also should have made his own efforts to reach the state of having high ambition, new ideas, and the power of invention.

6.2.5 Restructured a Department with My Own Thoughts and Failed – Resources 2

Event: To keep an independent department profitable, I tried to lower the payment to the headquarters by cutting staff members but the attempt failed.

Background: The drop in sales quantities and prices were pointing towards a business loss for our department and I wanted to find ways out of the situation (Figure 6.20).

Mind process: Concerned with possible business loss, I was looking at the profit and loss statement when suddenly I got an idea: "Gross profit will not improve so quickly, but it is still not in the red. If I can reduce the headquarters fee (and it was US$12M!), then I could probably keep the department out of loss."

The headquarters fee was based on the headcount in the department. It was about US$50K per head and if, for example, I sent 40 employees to another related company then I could save 40 x 50K = 200K.

Sending employees from our department to subsidiaries for the reorganizing and greater efficiency of our operation would also contribute to reducing the headquarters fee. I thought I had come across a great idea. I quickly built a large-scale reorganization plan, had the union agree to it, and the employees as well, and swiftly executed the plan before the next fiscal year began.

Results: When the new fiscal year started, the headquarters sent me the profit and loss statement, and to my surprise, it showed even a larger headquarters fee. Stunned by the statement, I started to analyze the reason and found that the cause of this failure was my crude self-satisfying misunderstanding.

If the total amount of the headquarters fee remained unchanged and if the total headcount for the entire corporate group changed, the headquarters fee per head would also change. If our department reduced the headcount while others added the same amount, then our headquarters fee would have changed as I expected (Figure 6.21). If other departments did not add new heads, then our reduction of the headcount would reduce the headquarters fee but not as much as expected. If other departments reduced the headcount at a rate greater than the cutback in our department, our headquarters fee would increase despite our reduction in the workforce, and this is what had happened.

Not understanding the basics of the headquarters fee and processing a plan based on my misunderstanding of "50K per head" resulted in my failure to reorganize our department.

Summary: To avoid a departmental financial loss, I tried to cut down the headquarters fee; however, other departments were conducting similar reorganizations and the outcome did not turn out the way I had expected.

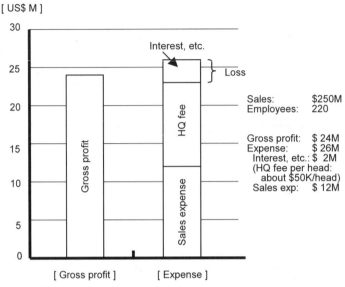

Figure 6.20   Gross profit and expense by the department

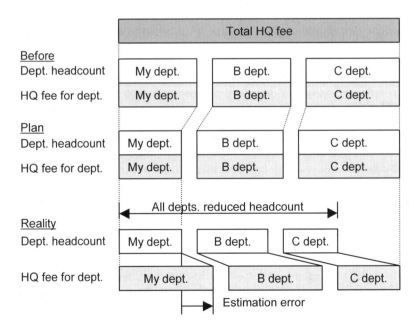

Figure 6.21   Plan for HQ fee reduction and what actually happened

Afterthoughts: It may sound like an excuse, but my headcount reduction was not just trying to match numbers. I divided the work into two groups: those that should remain in our group, and others that we could transfer to subsidiaries for better overall efficiency. It was in the right direction in the long run.

Failure just came from a greedy mind that tried to make the short-term results look good by taking advantage of the headquarters fee.

Knowledge:
- When you get a good idea, others may well have the same one.
- Quick ways of earning are not lying around.
- The headquarters fee can only go down through trimming the headquarters.
- Playing with numbers leads nowhere

6.2.6 Started a Young Engineers Training Program – Management 1

Event: In October 1996, our company started a project team for planning events for the tenth anniversary of our corporate establishment. One of the projects was an "Event for Enhancing the Level of Young Engineers." Four young group managers gathered and had a number of sessions to come up with a proposal to the board meeting for a "Corporate Research Presentation Seminar." The board accepted the overall idea, but said, "Reevaluate the details" and was hesitant in giving us the final go-ahead. Over the next 6 months, we repeated the proposal several times and through our work with the board its contents gradually made improvements, so that at the end a plan totally different in its details from the first one was accepted and executed in July 1997. It was named the "Young Engineers Forum;" it was the result of pushing the original idea up through the spiral of improvements and it is now an established event with a high level of practicality in the company.

Course: The original proposal linked the concepts of "activate young engineers" and "increase the opportunities of presentations;" at the same time, the idea of "interactive informational communications among multiple departments" was in the background of the proposal. We interactively correlated and analyzed the relations of "internal vs. external" and "short-term vs. medium to long-term" for the event. We proposed a "Corporation-wide research presentation meeting" aiming at "internal" and "medium to long-term" effects. We based our proposal on the fact that another event had a "private fair," an "annual publication," and a "corporate showroom" (Figure 6.22). Our concepts at the time were "gathering," "interaction among different business units," and "a place for free discussion."

The board, however, after listening to our presentation said, "The overall basic concept is good, but the research presentations tend to be boring. We want to learn more about real technology from the outside." We then conducted hearings with individual departments for the second time. Each department strongly opposed our idea as well. For while we had assumed the research presentations would be for reporting accomplishments and for sharing information and opinions, the departments were interested in an opportunity to appeal to individuals and groups. We then had to remove the original concept of activating young engineers from the first priority.

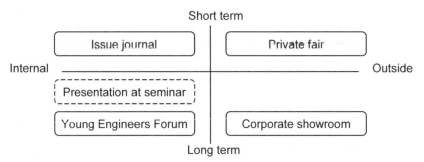

Figure 6.22   Relations among projects for the anniversary event

Our investigations resumed. Employees aged around 30 were the main force of the company, and we started interviewing them about their needs. It turned out that most of them complained about "insufficient time," followed by "co-workers," "boss," and "wages." We then took the suggestion from the executives of "interaction with the outside," combined it with results from the employees of "time, allowance, and co-worker" as the key phrases, and developed a new central concept of, "opportunities to interact with different industries." We also added "group independent research" and "opportunity for an appeal" to the main concept, and once more submitted the plan again to the board.

The result was another rejection with the comment that "interaction with different industries would not last long without good common grounds." During our next reinvestigation, we came up with the important key phrases of "recognition," "creation under unusual circumstances," and "pursue oneself". The company, however, already had its schemes of "venture business proposal" and "team improvement proposal" and our proposal would end up being an insignificant one. Then with the members of the proposal team getting busy with their work, the new event reached a deadlock.

Decision: One of us suggested removing the concept of "training," which tends to be a top-down assignment, and restructure the event with concepts of "autonomy," "consent," and "voluntary." In fact, "activating young engineers" should not be pressed from the top, and "autonomy" was the most important part. We then concluded that instead of the subjective word "training" we should use the voluntary word "forum." This led us to advance the original idea of having the administration staff manage the event based on the idea that "engineering managers will manage the event and administration staff will provide support." The ideas of "corporate-wide research presentation meeting" or "communication among multiple disciplines," that we came up with during our earlier plans, would not be suitable for the final purpose, but we figured they would make good additions as part of the whole menu. We titled the entire proposal to "The Young Engineers Forum" and presented it to the board meeting for the third time.

The board gave positive comments like, "It is good because human networks that one builds himself will last well," "The ideas for continuation will keep the event growing and engineering managers are better suited to that purpose," or "You will need ways to follow up even after the participants return to their jobs."

The proposal was accepted. Our half-year-long discussion led to the forum. It started with an internal advertisement for participating in "The First Young Engineers Forum" a year later. Some of the participants have made it to the management side, and other mechanisms for continuing and progressing the event have been adopted. The event improves itself every year and the fourth Young Engineers Forum is underway.

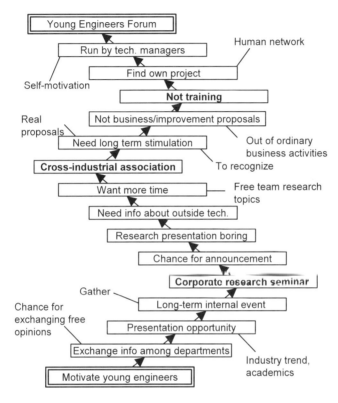

Figure 6.23   Relations among projects for the anniversary event

Mind process: Figure 6.23 shows how the original concept of "activating young engineers" moved up the spiral of the relation of thoughts diagram. The figure shows how the original concept deviated from the real purpose with constraints from other annual events and how it was corrected, how it distinguishes itself from one-time events like communication among multiple disciplines, and how it turned into a voluntary event by discretely stimulating areas that are hard for the engineer to recognize through his daily work. Figure 6.24 displays the "mind of the participants"

Knowledge: When we broke down the original large target, we misunderstood the constraints, which in the total concept being accepted but the details being rejected. Planning is the act of the requester and planner going back and forth to clarify the uncertain constraints.

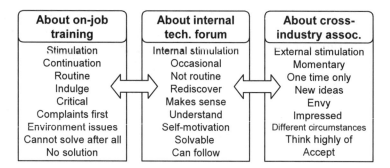

Figure 6.24 Differences in the minds of participants

The "explanation tool" and "presentation skills" are important for the engineer in outputting his ideas. They change the "margin of the mind" and "intellectual stimulation." Data of the information transfers well; however, transferring the mind of the information is difficult. From my experience, the mind transfers best after 9 p.m. on Fridays.

I sense an element of high corporate pride in this article. Placing the focus on "autonomy" at the end shows the high potential of a Japanese company; however, the individuality of employees should have come out more strongly. Autonomy within the closed world of a company may lead to merely the surface of the event being presented in later years.

6.2.7 Applied to a MITI Project but I Was Declined – Management 2

Event: MITI had asked us if we wanted to apply for one of their national projects to jointly pursue a specific research, but we had turned the offer down. Six months later we suddenly changed our minds and filed an application but was rejected.

Background: At the time, we started research into applying the newly developed technology for environmental protection to our model. Soon after we started the research it was the subject of quite a big article in a newspaper. Our group members would ask, "Who leaked the information?" but we never thought of any further action. At the time environmental protection was the trend and the Ministry of International Trade and Industry (MITI) also wanted to take part in this topic. To our advantage, MITI requested us to provide cooperation and participation in their project because the aforementioned newspaper article left the impression that we were leading the technology.

Discussion 1: We evaluated whether we should participate in the project or not as Figure 6.25 shows.

First of all, participation would promise plenty of research funding from the government. The challenging project theme would promote our technical ability to the market, and morale among our engineers would go up.

On the other hand, the disadvantages of the national project would be the difficulty in changing the project theme in the middle, and that once started, we would be overwhelmed with having to record our daily activities and submitting reports, that the accomplishments would be owned by the government, making it difficult to keep secrets, that the market needs were not well understood at the time, and that it would be better to wait for the element technologies to mature.

At the end, because there was no other manufacturer that could pursue the research theme, and since even if a competitor caught up with us it would not be too late for us to discuss our action then, we decided not to participate in the project.

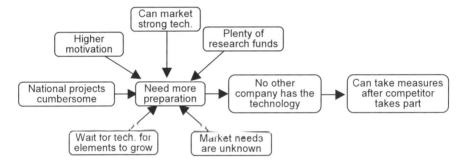

Figure 6.25   Expansion of thoughts about project participation (Part 1)

Decision 1: We politely declined participation in the project; sinceMITI and our competitors never dreamed of our rejection, they then thought that we were very close to putting the technology into mass production. Our rejection had caused them to put a sense of urgency into the proposed project. On the other hand, after we made the decision, the resources for the project were cut back and the person in charge had to run around trying not to let the project die.

Change in condition: But the world is as kind as it is cruel. After continuing the research quietly, we discovered that our competitor A was in the project. Also, at the same time, our director changed to a new person who was positive about the project.

Discussion 2: Figure 6.26 shows how the above changes made us think differently about the national project.   With our competitor in the picture, we thought we were faced with the following problems.
  (1) If the development budget went to our competitor, it would cause problems.
  (2) If our competitor beat us in the competition, it would injure our prestige and our having rejected the participation earlier would be raised into an issue.

(3) The rejection would leave an impression that our company was unsociable and might lead to difficulty with other projects (nobody might want to work with us ever again).

The change of director led us to believe that the market need for the technology was strong in both sales and service, and it was an important theme in terms of giving the market an image of our strong technology.

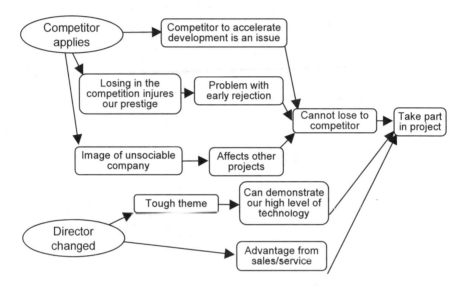

Figure 6.26   Expansion of thoughts about project participation (Part 2)

Decision 2: From the above reasons, we reversed our direction and decided to put our full efforts into applying for the project so that we would not lose to our competition.

Results: Our hope was turned down and we lost the competition to our competitor. We had to allocate our own budget so that we would not be left behind our competitor A. We were forced to limit our topics to within the budget and proceeded.

Afterthoughts: From our confidence of being the market leader, we decided whether to go forward or not only from our own viewpoint. We totally lacked the virtual exercise of what to do if a competitor applied for the project. Pride goes before a fall.

Whether it is an internal project or a public one, we should exchange information with the sponsor from the beginning about selecting topics and hear the plans for budget distribution. We should have carried out preliminary discussions and tests, so that we could have had what we wanted to do stated by the sponsor.

7

Applying the Mind Activity for Manufacturing

7.1 Where in Manufacturing to Apply the Mind Activity

This book primarily deals with design; however, once you understand the decision process, it is not just design that changes the methodology. Product planning, contents, and manufacturing methods also change. Not just the manufacturing contents and methods change, but also the speed changes. Once the speed changes, the cost also changes. The product then gains more margins and its market coverage expands. The profit, of course, also changes. Once we understand the decision process, the way products are made may change completely.

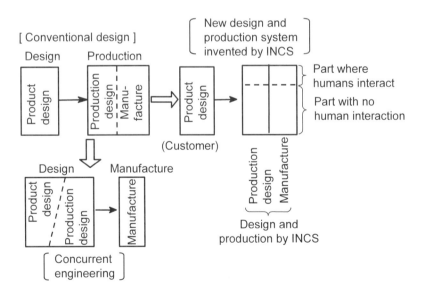

Figure 7.1 Dividing design and manufacturing work based on human interaction

Let's break down the process of manufacturing (Figure 7.1). Conventionally, we used to split design and manufacturing. Someone thinks and documents the product, then once the design is complete, passes it to production where the production engineer executes the manufacturing processes. The timetable starts manufacturing once design is finished. This conventional method lacks coordination, and the engineers came up with the concept of concurrent engineering (the idea of design and manufacture taking place at the same time) to plan design and production at the same time.

Another view distinguishes the man and the machine. This idea separates the method of production (production system) from the human beings. It draws a line between what humans think in deciding, and where the production system simply executes what has been decided. This distinction requires the clarification of "what and how" humans decide.

This idea allows having design and production at the same time and in same place. Clarifying "what the human decides" allows the development of a totally new production system. Such a line of thought was nothing but pie in the sky five years ago. There is now, however, a corporation that has actually implemented this practice and has achieved incredible productivity with precision products. One such example is INCS, inc. The production system at INCS has clearly distinguished what humans do and what machines execute, and then has completely rationalized the human decision part. INCS now produces prototype moulds (Figure 7.2) for cellular phones all around the world and has dominated the market. It produces a prototype within 1/5 to 1/10 of the time it used to take.

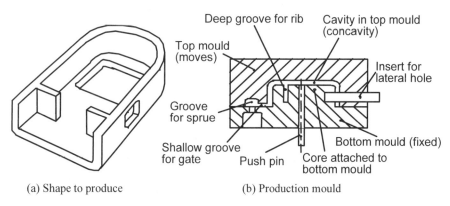

(a) Shape to produce   (b) Production mould

Figure 7.2   Moulds for cellular phone production

7.2 Decisions by INCS

How did INCS plan design and production? Whether it is design or production, they split the processes into parts where humans interact and those that machines can handle once humans have made the decisions. Such planning was not possible when the machine systems had not quite been digitalized and required human interaction, but now it is the actual digitalization of machine systems and information systems that makes it possible to separate human mind from the machine system. Beyond the level of CNC, all machines are computer-controlled and all processes of manufacturing are detectible with sensors, while all hardware that used to require human interaction is now completely computerized. Once the instructions are given, the system conducts production without human interaction (manless manufacturing). So there is complete separation of the human part and the machine part; once a human decides what to produce and how, the machine can handle the rest. To implement such new engineering, however, required some new technologies (Figure 7.3).

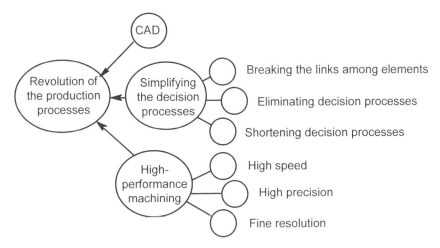

Figure 7.3 Revolution of the mould processes by INCS

First, they had to make a new CAD system. Next came a breakdown of the processes humans use to think. And lastly, a new high-performance machining method was required. These three, hardware, software, and system combined with observing what goes on in the human mind, made it possible to develop a totally new production system. We next describe this new system.

First they made a new CAD system. This system was totally different from a surface modeler or solid modeler. The new system may be called a grain modeler (Figure 7.4); it models the entire volume of a solid with spheres and each sphere carry all the attributes to define the part inside its space. Unlike those for a conventional solid modeler, the data do not belong to space coordinates; the spheres with their own volumes occupy the space and the intersection of two surfaces is always consistent with the model. This keeps valid data without discontinuity when, *e.g.*, calculating the cutter path and is convenient for the actual manufacturing.

Next, for the breakdown of the decision processes, they carefully observed the human operations to discover the detail of the human judgment and decision then uncovered the tacit knowledge. When there was a process that required human interaction, instead of taking the whole process as one that required human interaction, they decomposed the process into elements to clarify for which ones the judgments or decisions were made (Figure 7.5). This analysis revealed that a typical process believed to require human interaction, if decomposed into say 10 elements, really required human judgment or decision for only 2 or 3 of them. INCS further split the processes into finer elements and applied the methods described later in this chapter.

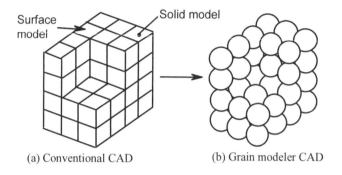

(a) Conventional CAD   (b) Grain modeler CAD

Figure 7.4   New CAD developed by INCS (KATACAD)

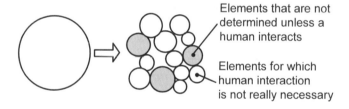

Elements that are not
determined unless a
human interacts

Elements for which
human interaction
is not really necessary

Figure 7.5   Extracting decision elements that require human interaction by decomposition

Lastly, for "high-performance machining," INCS realized high speed, high precision and fine resolution. By developing these three new technologies of CAD, human mind analysis, and high-performance machining, INCS combined the human work and unmanned machine work in a clear manner and first succeeded in separating the human and the machine.

7.3 Implementation by INCS

INCS started by building absolutely strong technologies. Not only did they achieve competitive pricing, but they also drastically enhanced precision and reduced the process time. For example, they reduced the manufacturing time not just by 20 or 30%, but targeted taking a digit off the time to 1/5 to 1/10 of the original. All manufacturing cost is proportional to the time it takes. In other words, if the time is decreased to 1/10, then the cost also falls to 1/10. Their thinking was that even if the hardware cost twice as much, reducing the manufacturing time to 1/10 of the original would cut the overall cost to about 1/5. The basic idea that INCS had was to reduce the time for each process. For realizing this idea, the new technologies of CAD, breaking down the decision processes, and high-performance manufacturing were needed.

First, we will evaluate the breakdown of decision processes. In shortening each process time, the biggest factor was the parts of the time spanned by human elements. The basic idea is to completely eliminate those parts sthat really did not need human interaction and, for those that did, to minimize the time. This analysis accomplished a drastic reduction in the process time.

Next we will discuss the basic idea of minimizing the human elements in manufacturing. First comes asynchronizing the relations among elements, which has two base methods, "parallel processing" and "asynchronization."

- "Parallel processing" means to carry out processes A, B, and C simultaneously without having to sequentially proceed process A, then B, and then C (Figure 7.6(a)).
- "Asynchronization" allows the machining of A, B, and C at arbitrary times of their own, by dissolving those relations among A, B, and C which used to require, for example, A, then B, and then C (Figure 7.6(b)).

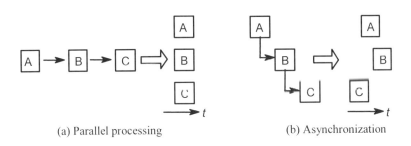

(a) Parallel processing   (b) Asynchronization

Figure 7.6   Removing relations among elements

The basic idea of these two analyses is removing the relation among elements. The next is "removing decision processes." Figure 7.7(a) shows this modification. Sometimes, people think they are making decisions, but in fact there is only one path to proceed on while we just hesitate or go back and forth. In this case, we should just move on without hesitation. This means removing what human beings thought were decision processes, whereas actually they were not.

Next is "removing options" (Figure 7.7(b)). If there is only one standard, there is nothing to wonder about, but there are often times when there appear to be multiple options. If these can be bound into a single selection, we can remove the questions when human make a selection. For example, if one process produces a continuous distribution with three peaks, we bundle them into three discrete lumps, and then instead of selecting one from the three peaks, define just one lump that can cover all three, and allow that selection. Also if there is a normal distribution, we cut off the two ends to bundle most of the distribution into one lump, and then force the selections to this one choice. We will later show an example that set the lengths of pins to a single short value that appled to all situations.

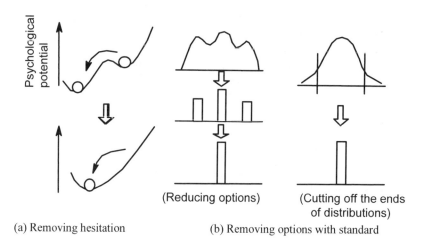

(Reducing options)   (Cutting off the ends of distributions)

(a) Removing hesitation   (b) Removing options with standard

Figure 7.7   Removing relations among elements

Next is "simplification." This analysis first requires "removing dependencies" (Figure 7.8(a)). It means to remove dependencies like "if A then B" as far as possible. Humans often mistake an independent relation for a dependency.

Next is "eliminating multiple preconditions." Say that process C only happens when process A and B are both complete. We may be able to change it so that we can proceed to C when only A is met and leave B alone (Figure 7.8(b)). An example is when we insert a key into a keyway. Normally we will measure the dimensions of the key and its keyway, but instead if we can constrain the key dimension to always stay within a tolerance, we only have to measure the keyway. If the keyway is finished to a dimension range we can advance to the next step (Figure 7.9).

"Minimizing the number of options" is also another form of simplification (Figure 7.8(c)). If we have to select from A, B, C, and D after some process, we apply the idea of removing dependencies in Figure 7.8(b) and say we have decided to no longer take A or D; we also make modifications if necessary so that if we choose either B or C then the required functions are met. Eliminating A and D in this case is a form of minimizing the number of options.

When INCS applied these ideas and produced a mould for a cellular phone casing, the actual process came out as Figure 7.10 shows. Using the CAD, the designers first determined the layout of cavities, core, shape, and pins. Then by removing interferences among the design processes, they were able to advance them in parallel (Figure 7.10(a)). Figure 7.10(b) shows the decision processes for the mould layout. Triangles indicate the start and end of works, and the long rectangle when the work is carried out. A diamond shape indicates a decision process that completes at the right end of the diamond. The breaking down of decision processes takes place at the beginning, using a CAD system. Next dependencies among decision process elements are split, then made parallel and asynchronous, and decision processes are deleted, all as far as possible. Then, simplifying the remaining decision process elements they are connected all along the time line. Figure 7.10(b) provides the resulting process diagram, showing a monotonic plan without any element going back and forth and the entire process connected in a single path from the left to the lower right. In other words the analysis removes interference and synchronization.

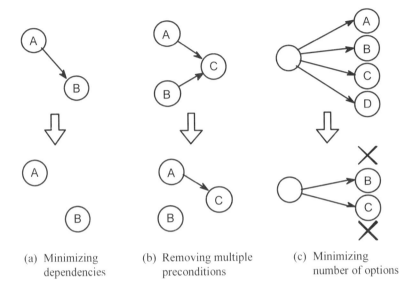

(a) Minimizing
     dependencies

(b) Removing multiple
     preconditions

(c) Minimizing
     number of options

Figure 7.8   Simplifying decision processes

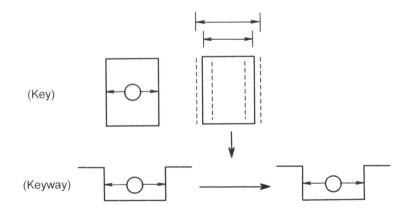

(Key)

(Keyway)

Figure 7.9   Example of simplifying the decision process

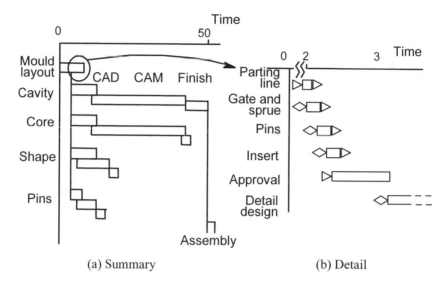

Figure 7.10   Sample process diagram of the revolutionized decision design process for the cellular phone shell

7.4 Outcome of the INCS Development

Conventionally, these moulds were fabricated by first machine-forming the cavities, cores, and overall shape. INCS then used EDM to cut the vertical walls and open complex holes. This process involved cutting out the negative image of the hole (male), then grinding to finish the surfaces. INCS, however, used their new CAD system, broke down the decision processes and developed high-performance machining to completely eliminate the EDM process and shape the mould just by cutting. Supported by these machining hardware developments, the time to produce this mould was drastically reduced.

First for machining, INCS developed a 50,000rpm miniature spindle (based on the high-speed spindles the dentists use). The process development further enabled fast acceleration and stopping of 2G for moving the spindle.

Next they succeeded in producing a long, fine end mill with diameter 0.4mm and depth 14mm. The tip oscillation was 2µm wide. This end mill could cut a truly vertical wall, so EDM was no longer needed.

Another accomplishment was to completely eliminate the need for finish grinding by measuring the end mill friction and thermal deformation.

Successful development of the hardware and the new grain modeler CAD (named "KATACAD") definitely simplified the human mind processes and shortened the CAD process (engineering) hours from about 64 hours down to about 1/10 or 6.5 hours. The hardware and rescheduling the human interaction also shortened the entire design, machining, assembly, and finishing process time from 352 hours (about 44 days) to 50 hours (about 6.2 days), which is about 14% of conventional time (Figure 7.11). These developments made it possible to meet the needs of the current market and INCS started to accept all the orders from around the world for prototype moulds of cellular phones.

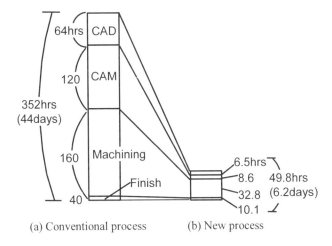

(a) Conventional process   (b) New process

Figure 7.11   Effect of INCS new mould production process

7.5 Where Does This Lead Us to in the Future?

INCS have accomplished astounding productivity by making use of the mind activity. What have they done so far and where are they going from here? INCS first succeeded in rapid prototyping (technology that turns small particles into 3D solids based on the 3D information) by treating it as a 3D printing technology, by realizing the technology and commercializing it. Then they looked at the mind activity for mass production of metal moulds. They are now in consultation with a number of manufacturers to review their production systems. They want eventually to enter the business of manufacturing education.

In the background of these developments lies the experience of Mr Shinjiro Yamada, the president of INCS. He worked for a long time on developing the door lock in an automobile part company, and attained comlete success in how to think and work the design. His accomplishment was sold to Chrysler and then to Honda. He thought, however, that there would be not much future advancement in the door lock design, left to start a business that dealt with a much bigger manufacturing system and started INCS, inc.

What is important here is that he completely mastered the fundamental technology of the hardware, planned the 3D CAD to use the technology, realized that the mind   activity is the essence of production activities, and proceeded to use these facts.

Many had the idea that making use of the thinking process is the most important factor in determining production, but no one could put the idea into a business. Here it was realized and this is the way we have made progress in our history.

Next, we will ponder about the transition of production, that is, what it relies on, and how the object changes. The history of mankind shows humans relying only on natural energy and materials for a long time. After the industrial revolution, people started to rely on power and energy. In the late twentieth century, we started to rely on information technology. From the twenty-first century, the root of production will be the area expressed in terms "knowledge" or "brain." The overall movement is shifting the basis of production along the sequence Hardware → Energy → Information → Knowledge. Unless we accept this fact, we will not be able to properly grasp the future of production.

So, what is the basic science when we think about the activities of "knowledge" and "brain"? We believe it is brain science. Currently we only watch and try to prove our theories about what goes on in the brain and explain what it is that happens in the brain. In the future, we will certainly make use of the knowledge we acquire this way and will aim at contributing it to human activities. It is the product of brain science and its contribution to the mankind. When we think along these lines, how we view production changes its appearance. As we saw in the example of INCS, production probably has three main pillars, namely, software, hardware, and system, or maybe the system has its roots within the mind   activities. This is exactly what Figure 7.3 claims. At the same time, CAD/CAM, which was merely a combination of CAD, which used to be an electronic drawing machine, and CAM, which was just a computerized NC, will grow into CAM/CAM in the true sense that connects 3D information and the mind   activities.

In August 2001, the author (Hatamura) attended a CIRP (International Institution for Production Engineering Research) meeting and had a chance to talk with M. Bernard Charlès, the President and CEO of Dassault Systèmes. IBM CATIA is a product base on Dassault's 3D CAD system. M. Charlès said that Dassault had about 2,000 engineers working the 3D CAD product and that in the year 2002, it planned to add another 1,000 to actively incorporate the mind activity of humans and strive to realize CAD in the real sense supporting the human mind. The author believes that this is the right direction.

The authors have advocated the "principle of creative design" that claims that there are patterns in human thinking and that making use of the patterns brings many benefits. Our claim has common grounds with TRIZ that Altshuller proposed in Russia over 50 years ago. It now has some commercialized computer software based on the idea. The principle of creative design actively supports the mind process of the humans involved with such tools as the "expansion of thoughts diagram", "hidden drawing", and "showing failure cases". These ways of thinking will probably be the most important methods for production for the next generation.

The current talk in Japanese industry points at overseas production and the de-industrialization of Japanese manufacturers. At the same time many voice the need for passing down the technologies to the next generation. The problems pointed out and what each individual says is correct. When we look at the solution to these problems, however, the author thinks the arguments miss the point. I believe that "active use of brain activities for production" will counter this overseas production problem to continue Japanese manufacturing and allow passing the technologies down further. I believe that mere discussion neglecting the active use of brain activities will lead nowhere.

The author (Hatamura) believes that Japan, other leading countries, and companies that want to use new technology and rely on it to survive will practice "putting the technology in a black box." The black box of technology means the "three noes" of "no showing", "no speaking", and "no touching".

First is not to show. If you show some technology to others, they know that it is a reachable point. When there is nothing, people do not know whether such things are attainable or not. Thus, they do not even know if struggling in that direction is justifiable or not. Once you show it to someone, you have just revealed that there is a solution and you have just built a lighthouse for the moonless night. This is the meaning of showing. Also, witnessing the reality, people receive a lot of stimulation that generates a flow of ideas and eventually they can make it reality. Showing a product is fine, but the production technology should never be opened.

Next is not to speak. Not to speak does not just mean not talking about what you saw or heard about. People that are struggling to succeed can instantly understand the whole route, once they hear about what hesitations others had, what options others tried, what went through others' minds, or what failure others experienced. The followers can easily trace the path you have worked so hard to build. And someday, the follower will beat the pioneers. Once you understand this, you will know it is a very bad idea to talk about the technological journey you made. That is why we should not speak.

Last is no touching. Once you touch it, you can understand it. You can let others touch and analyze a product, however, letting them touch the route you traveled would let them understand the technology. We should not let others touch the ways of our production. Why are these "three noes" of "no showing", "no speaking", and "no touching" not valued in Japan? It is because Japan only has experience in following success cases in the western world and these efforts have always been rewarded. Any company in the western world plays the game on the principle of the "three noes," especially if the technology has been developed by themselves. When we discuss technology with members of such companies, we feel the huge ideology of technology they have behind them. We will establish the black box of technology to cover things both visible and invisible. An important thing to note here is that "it is impossible to steal original culture." Each country, company, or person has to give birth to the technology from within.

Postscript

It took us six years to complete this book. Book 1, *The Practice of Machine Design – Theory and Methods of Machine Design* in the "Practice of Machine Design" series was published in July 1988, and Book 2, *The Practice of Machine Design – Knowledge and Data for Machine Design* in July 1992, and Book 3, *The Practice of Machine Design – Learning from Failure* in October 1996. This fourth book is titled *The Practice of Machine Design – This is How I Decided.*

    Note: The above paragraph refers to the Japanese versions. The corresponding English versions are:

- *The Practice of Machine Design*, Oxford University Press, 1999 This English book combines Book 1 and Book 2 of the original Japanese version.
- *The Practice of Machine Design – Learning from Failure*, TBP
- *Decision-Making in Engineering Design*, This book is the translation of the fourth book in Japanese.

We edited this book to clarify what every designer wants to know: how each matter was determined, in what order, what the constraints were, what could change and what could not, what were the givens, what were the relations among these matters, and so on (This information is usually not explicitly written down on the drawings.)

We started on the book immediately after publishing Book 3, and as always proceeded at the rate of having six meetings a year, each participant submitting his article proposal, and everyone discussing them. It took us forever, however, to complete the manuscript for *This is How I Decided* and the scheduled Olympics year of 2000 went by. After thinking about what caused this delay, we concluded that we had tried too hard to produce something society was looking for and so in the fall of 2001 we went back to our original intent of "produce what we want to", and this is the book that came out.

When the editor in chief (Hatamura) was a senior in college, there existed the "graduation assignment" (perhaps it was an assignment for the junior students). One of the choices was to design an engine. His colleagues read many books, which taught them, this was how normally done and they just copied them on their drawings, but Hatamura was never satisfied with that method. Especially for designing engines, the book by Professor Fujio Nagao, then of Kyoto University, had plenty of information in it. All these books, however, lacked an account of the design process, describing what was variable and what was not, what was determined by technical constraints and how they were set. Since then, he has therefore, wanted to produce someday a book that described all these traces and how things were decided. He thought such a book would help many of us in our studies, and after 40 long years, we have finally come up with the book.

We have written on various topics in this book. Hatamura sees design as an act of "proving a hypothesis" about a proposed solution to meet functional requirements under given constraints, and he believes brain science will support this theory. Researches by Dr Gen Matsumoto of RIKEN and others will lead us to directly watch the activity of the brain and it seems that this will prove that the route that is electrically excited within the brain is what we have described in this book.

Let's review the activities by us, the Practice of Machine Design Research Group, and how the activities relate to this book. Many of the members of the Practice of Machine Design Research Group joined the "Creative Design Engine" project that started 4 years ago. The Japanese Ministry of International Trade and Industry supported the project and invested about US$600,000 a year for three years. The project aimed at realizing a system that supported the natural mind process of a person working on a creative project like design. The project was completed with a prototype. From the year 2000, with the support of the Ministry of Education, Culture, Sports, Science and Technology (MEXT), a new project started with the "Failure Knowledge Research Group." We have been working to assemble a "Failure Knowledge Database" for this project since 2001 and will continue for about 5 years. This project also promotes the display of understandable cases of failure knowledge whenever designers and project planning engineers require it. In addition, since December 2002, the authors have been working to establish the "Association for the Study of Failure." This group also works to uncover the tacit information in design, *i.e.*, how someone thought and how they failed. It also conducts research on failures themselves as well as on how to make positive use of them so that the designer can think about technology as well as other things going on in our society. We hope the reader will enjoy all our books.

2002, On behalf of all the authors, Yotaro Hatamura

Index

action, 17
age and abilities, 217
age of 30, 204
anxiety, 8
archive, 18
Assistant, 210
association, 18
asynchronization, 253
automatic grinding, 90
automatic segment construction
robot, 137
automobile compressor, 122
background, 18
backhoe, 131
balance weight, 122
breaking rocks, 195
bridge over the pacific, 218
CAD, 50, 251
3D CAD, 189
KATACAD, 256
CAE, 51
CAM, 192
3D CAM, 192
cause, 17
centrifugation, 177
China, 16
clamshell digger, 131
cluster, 5
company to school, 210
compliance, 90
negative compliance, 91
constraint, 1

constraints, 18
consulting, 207
corporate management, 231
crane, 142
Creation of Technology Research
Group, 230
creative design, 234
Creative design engine, 98
decision, 7, 17, 24
decision-making, 2
make a decision, 2
types of decision, 2
Department of Engineering
Synthesis, 233
design
axiomatic, 127
contents of design, 2
creative design, 40
design documents, 46
design manual, 39
design record, 46
design support system, 49
former design, 39
Development, 113
Diamond Princess, 11
die
injection molding die, 158
disabled, 221
discussion, 18
distribute, 18
DNA, 171
DNA anchoring, 171

DNA fragmentation, 176
doctoral thesis, 211
doubt, 7
doubts, 18
educate, 19
electromagnetic force, 195
element relation, 26
engineering doctor, 225
evaluation, 18
event, 17
Failure Knowledge Research Group,
    262
final plan drawing, 42
fine positioning in vacuum, 52
flatterer, 213
fluorescent molecule, 176
follower, 39
force flow diagram, 63
forming with lateral vibration, 198
function, 40
functional requirement, 1, 18, 40
galvanometer, 186
go or no go, 7
hidden drawing, 49
hidden image mirror, 153
high-precision positioning, 185
hydraulic cylinder, 56
Hydraulic cylinder, 53
hydrolyzation, 36
idea notes, 15
INCS, inc., 250
intelligent grinding machine, 82
investment, 236
irreversibility of time, 20
junction hood, 127
Kintaro-candy, 185
knowledge, 17
learning, 5
lighting, 150
make business, 19
make selection, 4
man and machine, 250
Mandala, vii, 26
manuals, 37
manufacturing, 45, 113
market, 45
material cooling system, 170

MEMS, 225
MEXT, 262
micro-machines, 86
microscopic assembly, 180
microscopic tweezers, 181
mind
    mind potential, 2
    mind process, 1
mind process, 21
    basic mind process, 41
miniature power source, 225
minimizing number of options, 254
MIT, 225, 227
MITI, 86, 246
MMM, 209
Modification, 113
motivation, 18
mountain climbing shoe, 20
N.P. Suh, 127
nano-manufacturing world (NMW),
    86
Narita Express, 127
national project, 247
NEDO, 226
occupation, 204
operation
    forward operation, 39
    reverse operation, 39
organizational change, 233
outrigger, 142
parallel processing, 253
peeling, 31
plane of thoughts, 5
plasma display panel (PDP), 161
polygon mirror insertion, 65
polymerization, 36
positioning table, 77
Practical Solution, 113
practice, 6
Practice, 113
pressure sensor, 118
Principle of Design, 127
procedure, 39
process, 17
proposal, 46
psychological obstacle, 9
publicize, 19

radio-frequency identification
   (RFID), 158
realization, 18
record, 14, 18
related events, 18
remote safety system, 114
removing options, 253
Research, 113
restructure, 241
results, 18
robot
   multi-joint robot, 164
route of thoughts, 18
route to decision, 3
sandblasting, 161
scenario, 11
sealing, 63
search, 18
segment, 137
selection, 24
Seventeen rules, 108
severe environment, 118
simplification, 254
single selection, 8
six-axis force sensor, 65
socialize, 19
spore diagram, 26
static charge, 180
Stirling engine, 102
store, 18
strain gauge, 66
straw shoes, 33
structural element correspondence,
   27
structural element transition, 26
structural selection, 8
structure, 40

successor, 239
summary, 17
Super-Kamiokande, 8
supply chain management, 218
System Introduction, 113
Technology Introduction, 113
technology management, 189
technology transfer, 13, 14
telescopic arm, 131
Thinking, 113
thoughts
   expansion of thoughts, 24
   fragments of thoughts, 5
   plane of thoughts, 5, 23
   relation of thoughts, 23
   route of thoughts, 5
   spiral of thoughts, 25
three no's, 259
time history progress, 25
tip distance, 154
torque sensor, 65
trading company, 212
training young engineers, 243
TRIZ, 9
tunnel machine, 137
turbine blade, 90
turning independent, 218
turning point, 214
underground construction, 131
unmanned machine, 116
virtual exercise, 7
virtual structure, 3
wafer character reader, 148
weight reduction, 122
wheelchair, 221
YAG laser welding, 154

Printing: Krips bv, Meppel
Binding: Stürtz, Würzburg